Introduction to Geodesy

Introduction to Geodesy

Edited by
Gloria Nisbet

Larsen & Keller
www.larsen-keller.com

Introduction to Geodesy
Edited by Gloria Nisbet
ISBN: 978-1-63549-682-6 (Hardback)

© 2018 Larsen & Keller

⊟ Larsen & Keller

Published by Larsen and Keller Education,
5 Penn Plaza,
19th Floor,
New York, NY 10001, USA

Cataloging-in-Publication Data

Introduction to geodesy / edited by Gloria Nisbet.
    p. cm.
Includes bibliographical references and index.
ISBN 978-1-63549-682-6
1. Geodesy. 2. Geomatics. 3. Geophysics. I. Nisbet, Gloria.
QB281 .I58 2018
526.1--dc23

For more information regarding Larsen and Keller Education and its products, please visit the publisher's website www.larsen-keller.com

Table of Contents

Preface

The study of Earth's gravitational field and its representation in a three-dimensional time-varying space by applying the rules of applied mathematics and Earth science is known as geodesy. It also includes the study of crustal motion, polar motion and the cause of tides. The methods used to measure these motions are gravimetry, geodetic astronomy, levelling, satellite geodesy, etc. This book elucidates the concepts and innovative models around prospective developments with respect to geodesy. Most of the topics introduced in it cover new techniques and the applications of the subject. For all those who are interested in the vast subject of geodesy, this textbook can prove to be an essential guide.

A foreword of all Chapters of the book is provided below:

Chapter 1 - Geodesy is a subject which studies the representation and measurement of planets. Crustal motion, polar motion and tides are some of the topics studied under the subject of geodesy. This chapter will provide an integrated understanding of geodesy; **Chapter 2 -** Gravity of the Earth can be denoted by the symbol g. The force of the gravity of Earth depends on it location. Some of the topics discussed related to the gravity of Earth are theoretical gravity, gravimetry, gravitation of the moon, gravimeter and gravity gradiometry. This chapter elucidates the crucial theories and principles of the Earth's gravity; **Chapter 3 -** Geodetic datum is a system that is used to locate places on the surface of the Earth. The other concepts related to geodesy are geoid, horizontal position representation, figure of the Earth, metres above sea level and geographical distance. The topics discussed in the chapter are of great importance to broaden the existing knowledge on geodesy; **Chapter 4 -** The technologies related to geodesy are satellite navigation, global positioning system, Lidar and interferometric synthetic aperture radar. The system that uses satellites to locate places on Earth is known as satellite navigation. BeiDou navigation satellite system, Galileo and GLONASS are some of the examples of satellite navigation. Tools and techniques are an important component of any field of study. The following chapter elucidates the various tools and techniques that are related to geodesy; **Chapter 5 -** The exact calculation of various locations on Earth require the study of the shape of the Earth. The Earth's figure is theoretically explained as an oblate ellipsoid due to its rotation. Geodesy is best understood in confluence with the major topics listed in the following chapter.

I would like to thank the entire editorial team who made sincere efforts for this book and my family who supported me in my efforts of working on this book. I take this opportunity to thank all those who have been a guiding force throughout my life.

Editor

An Overview of Geodesy

Geodesy is a subject which studies the representation and measurement of planets. Crustal motion, polar motion and tides are some of the topics studied under the subject of geodesy. This chapter will provide an integrated understanding of geodesy.

Geodesy

An old geodetic pillar (1855) at Ostend, Belgium

Geodesy also known as geodetics, geodetic engineering or geodetics engineering — a branch of applied mathematics and earth sciences, is the scientific discipline that deals with the measurement and representation of the Earth (or any planet), including its gravitational field, in a three-dimensional time-varying space. Geodesists also study geodynamical phenomena such as crustal motion, tides, and polar motion. For this they design global and national control networks, using space and terrestrial techniques while relying on datums and coordinate systems.

Definition

Geodesy — from the Ancient Greek word *geodaisia* (literally, "division of the Earth") is primarily concerned with positioning within the temporally varying gravity field. Geodesy in the German-speaking world is divided into "higher geodesy", which is concerned with measuring the Earth on the global scale, and "practical geodesy" or "engineering geodesy" ("Ingenieurgeodäsie"), which is concerned with measuring specific parts or regions of the Earth, and which includes surveying. Such "geodetic" operations are also applied to other astronomical bodies in the solar system. It is also the science of measuring and understanding the earth's geometric shape, orientation in space and gravity field.

The shape of the Earth is to a large extent the result of its rotation, which causes its equatorial bulge, and the competition of geological processes such as the collision of plates and of volcanism, resisted by the Earth's gravity field. This applies to the solid surface, the liquid surface (dynamic sea surface topography) and the Earth's atmosphere. For this reason, the study of the Earth's gravity field is called physical geodesy by some.

Geoid and Reference Ellipsoid

The geoid is essentially the figure of the Earth abstracted from its topographical features. It is an idealized equilibrium surface of sea water, the mean sea level surface in the absence of currents, air pressure variations etc. and continued under the continental masses. The geoid, unlike the reference ellipsoid, is irregular and too complicated to serve as the computational surface on which to solve geometrical problems like point positioning. The geometrical separation between the geoid and the reference ellipsoid is called the geoidal undulation. It varies globally between ±110 m, when referred to the GRS 80 ellipsoid.

A reference ellipsoid, customarily chosen to be the same size (volume) as the geoid, is described by its semi-major axis (equatorial radius) a and flattening f. The quantity $f = (a - b)/a$, where b is the semi-minor axis (polar radius), is a purely geometrical one. The mechanical ellipticity of the Earth (dynamical flattening, symbol J_2) can be determined to high precision by observation of satellite orbit perturbations. Its relationship with the geometrical flattening is indirect. The relationship depends on the internal density distribution, or, in simplest terms, the degree of central concentration of mass.

The 1980 Geodetic Reference System (GRS 80) posited a 6,378,137 m semi-major axis and a 1:298.257 flattening. This system was adopted at the XVII General Assembly of the International Union of Geodesy and Geophysics (IUGG). It is essentially the basis for geodetic positioning by the Global Positioning System and is thus also in widespread use outside the geodetic community.

The numerous other systems which have been used by diverse countries for their maps and charts are gradually dropping out of use as more and more countries move to global, geocentric reference systems using the GRS 80 reference ellipsoid.

Coordinate Systems in Space

The locations of points in three-dimensional space are most conveniently described by three cartesian or rectangular coordinates, X, Y and Z. Since the advent of satellite positioning, such coordinate systems are typically geocentric: the Z-axis is aligned with the Earth's (conventional or instantaneous) rotation axis.

Prior to the era of satellite geodesy, the coordinate systems associated with a geodetic datum attempted to be geocentric, but their origins differed from the geocentre by hundreds of metres, due to regional deviations in the direction of the plumbline (vertical). These regional geodetic data, such as ED 50 (European Datum 1950) or NAD 27 (North American Datum 1927) have ellipsoids associated with them that are regional 'best fits' to the geoids within their areas of validity, minimising the deflections of the vertical over these areas.

It is only because GPS satellites orbit about the geocentre, that this point becomes naturally the origin of a coordinate system defined by satellite geodetic means, as the satellite positions in space are themselves computed in such a system.

Geocentric coordinate systems used in geodesy can be divided naturally into two classes:

1. Inertial reference systems, where the coordinate axes retain their orientation relative to the fixed stars, or equivalently, to the rotation axes of ideal gyroscopes; the X-axis points to the vernal equinox

2. Co-rotating, also ECEF ("Earth Centred, Earth Fixed"), where the axes are attached to the solid body of the Earth. The X-axis lies within the Greenwich observatory's meridian plane.

The coordinate transformation between these two systems is described to good approximation by (apparent) sidereal time, which takes into account variations in the Earth's axial rotation (length-of-day variations). A more accurate description also takes polar motion into account, a phenomenon closely monitored by geodesists.

Coordinate Systems in the Plane

In surveying and mapping, important fields of application of geodesy, two general types of coordinate systems are used in the plane:

1. Plano-polar, in which points in a plane are defined by a distance s from a specified point along a ray having a specified direction a with respect to a base line or axis;

2. Rectangular, points are defined by distances from two perpendicular axes called x and y. It is geodetic practice—contrary to the mathematical convention—to let the x-axis point to the north and the y-axis to the east.

Rectangular coordinates in the plane can be used intuitively with respect to one's current location, in which case the x-axis will point to the local north. More formally, such coordinates can be obtained from three-dimensional coordinates using the artifice of a map projection. It is *not* possible to map the curved surface of the Earth onto a flat map surface without deformation. The compromise most often chosen—called a conformal projection—preserves angles and length ratios, so that small circles are mapped as small circles and small squares as squares.

An example of such a projection is UTM (Universal Transverse Mercator). Within the map plane, we have rectangular coordinates x and y. In this case the north direction used for reference is the *map* north, not the *local* north. The difference between the two is called meridian convergence.

It is easy enough to "translate" between polar and rectangular coordinates in the plane: let, as above, direction and distance be a and s respectively, then we have

$$x = s \cos \alpha$$
$$y = s \sin \alpha$$

The reverse transformation is given by:

$$s = \sqrt{x^2 + y^2}$$
$$\alpha = \arctan \frac{y}{x}.$$

Heights

In geodesy, point or terrain *heights* are "above sea level", an irregular, physically defined surface. Therefore, a height should ideally *not* be referred to as a coordinate. It is more like a physical quantity, and though it can be tempting to treat height as the vertical coordinate z, in addition to the horizontal coordinates x and y, and though this actually is a good approximation of physical reality in small areas, it quickly becomes invalid for regional considerations.

Heights come in the following variants:

1. Orthometric heights

2. Normal heights

3. Geopotential heights

Each has its advantages and disadvantages. Both orthometric and normal heights are heights in metres above sea level, whereas geopotential numbers are measures of potential energy (unit: $m^2\ s^{-2}$) and not metric. Orthometric and normal heights differ in the precise way in which mean sea level is conceptually continued under the continental masses. The reference surface for orthometric heights is the geoid, an equipotential surface approximating mean sea level.

None of these heights is in any way related to geodetic or ellipsoidial heights, which express the height of a point above the reference ellipsoid. Satellite positioning receivers typically provide ellipsoidal heights, unless they are fitted with special conversion software based on a model of the geoid.

Geodetic Data

Because geodetic point coordinates (and heights) are always obtained in a system that has been constructed itself using real observations, geodesists introduce the concept of a *geodetic datum*: a physical realization of a coordinate system used for describing point locations. The realization is the result of *choosing* conventional coordinate values for one or more *datum points*.

In the case of height data, it suffices to choose *one* datum point: the reference bench mark, typically a tide gauge at the shore. Thus we have vertical data like the NAP (Normaal Amsterdams Peil), the North American Vertical Datum 1988 (NAVD 88), the Kronstadt datum, the Trieste datum, and so on.

In case of plane or spatial coordinates, we typically need several datum points. A regional, ellipsoidal datum like ED 50 can be fixed by prescribing the undulation of the geoid and the deflection of the vertical in *one* datum point, in this case the Helmert Tower in Potsdam. However, an overdetermined ensemble of datum points can also be used.

Changing the coordinates of a point set referring to one datum, so to make them refer to another datum, is called a *datum transformation*. In the case of vertical data, this consists of simply adding a constant shift to all height values. In the case of plane or spatial coordinates, datum transformation takes the form of a similarity or *Helmert transformation*, consisting of a rotation and scaling operation in addition to a simple translation. In the plane, a Helmert transformation has four parameters; in space, seven.

In the abstract, a coordinate system as used in mathematics and geodesy is, e.g., in ISO terminology, referred to as a *coordinate system*. International geodetic organizations like the IERS (International Earth Rotation and Reference Systems Service) speak of a *reference system*.

When these coordinates are realized by choosing datum points and fixing a geodetic datum, ISO uses the terminology *coordinate reference system*, while IERS speaks of a *reference frame*. A datum transformation again is referred to by ISO as a *coordinate transformation*. (ISO 19111: Spatial referencing by coordinates).

Point Positioning

Geodetic Control Mark (example of a deep benchmark)

Point positioning is the determination of the coordinates of a point on land, at sea, or in space with respect to a coordinate system. Point position is solved by computation from measurements linking the known positions of terrestrial or extraterrestrial points with the unknown terrestrial position. This may involve transformations between or among astronomical and terrestrial coordinate systems.

The known points used for point positioning can be triangulation points of a higher order network, or GPS satellites.

Traditionally, a hierarchy of networks has been built to allow point positioning within a country. Highest in the hierarchy were triangulation networks. These were densified into networks of traverses (polygons), into which local mapping surveying measurements, usually with measuring tape, corner prism and the familiar red and white poles, are tied.

Nowadays all but special measurements (e.g., underground or high precision engineering measurements) are performed with GPS. The higher order networks are measured with static GPS, using differential measurement to determine vectors between terrestrial points. These vectors are then adjusted in traditional network fashion. A global polyhedron of permanently operating GPS stations under the auspices of the IERS is used to define a single global, geocentric reference frame which serves as the "zero order" global reference to which national measurements are attached.

For surveying mappings, frequently Real Time Kinematic GPS is employed, tying in the unknown points with known terrestrial points close by in real time.

One purpose of point positioning is the provision of known points for mapping measurements, also known as (horizontal and vertical) control. In every country, thousands of such known points exist and are normally documented by the national mapping agencies. Surveyors involved in real estate and insurance will use these to tie their local measurements to.

Geodetic Problems

In geometric geodesy, two standard problems exist:

First (Direct) Geodetic Problem

Given a point (in terms of its coordinates) and the direction (azimuth) and distance from that point to a second point, determine (the coordinates of) that second point.

Second (Inverse) Geodetic Problem

Given two points, determine the azimuth and length of the line (straight line, arc or geodesic) that connects them.

In the case of plane geometry (valid for small areas on the Earth's surface) the solutions to both problems reduce to simple trigonometry. On the sphere, the solution is significantly more complex, e.g., in the inverse problem the azimuths will differ between the two end points of the connecting great circle, arc, i.e. the geodesic.

On the ellipsoid of revolution, geodesics may be written in terms of elliptic integrals, which are usually evaluated in terms of a series expansion; for example.

In the general case, the solution is called the geodesic for the surface considered. The differential equations for the geodesic can be solved numerically.

Geodetic Observational Concepts

Here we define some basic observational concepts, like angles and coordinates, defined in geodesy (and astronomy as well), mostly from the viewpoint of the local observer.

- The *plumbline* or *vertical* is the direction of local gravity, or the line that results by following it.

- The *zenith* is the point on the celestial sphere where the direction of the gravity vector in a point, extended upwards, intersects it. More correct is to call it a *direction* rather than a point.

- The *nadir* is the opposite point (or rather, direction), where the direction of gravity extended downward intersects the (invisible) celestial sphere.

- The celestial *horizon* is a plane perpendicular to a point's gravity vector.

- *Azimuth* is the direction angle within the plane of the horizon, typically counted clockwise from the north (in geodesy and astronomy) or south (in France).

- *Elevation* is the angular height of an object above the horizon, Alternatively zenith distance, being equal to 90 degrees minus elevation.

- *Local topocentric coordinates* are azimuth (direction angle within the plane of the horizon) and elevation angle (or zenith angle) and distance.

- The north *celestial pole* is the extension of the Earth's (precessing and nutating) instantaneous spin axis extended Northward to intersect the celestial sphere. (Similarly for the south celestial pole.)

- The *celestial equator* is the intersection of the (instantaneous) Earth equatorial plane with the celestial sphere.

- A *meridian plane* is any plane perpendicular to the celestial equator and containing the celestial poles.

- The *local meridian* is the plane containing the direction to the zenith and the direction to the celestial pole.

Geodetic Measurements

The level is used for determining height differences and height reference systems, commonly referred to mean sea level. The traditional spirit level produces these practically most useful heights above sea level directly; the more economical use of GPS instruments for height determination requires precise knowledge of the figure of the geoid, as GPS only gives heights above the GRS80 reference ellipsoid. As geoid knowledge accumulates, one may expect use of GPS heighting to spread.

The theodolite is used to measure horizontal and vertical angles to target points. These angles are referred to the local vertical. The tacheometer additionally determines, electronically or electro-optically, the distance to target, and is highly automated to even robotic in its operations. The method of free station position is widely used.

For local detail surveys, tacheometers are commonly employed although the old-fashioned rectangular technique using angle prism and steel tape is still an inexpensive alternative. Real-time kinematic (RTK) GPS techniques are used as well. Data collected are tagged and recorded digitally for entry into a Geographic Information System (GIS) database.

Geodetic GPS receivers produce directly three-dimensional coordinates in a geocentric coordinate frame. Such a frame is, e.g., WGS84, or the frames that are regularly produced and published by the International Earth Rotation and Reference Systems Service (IERS).

GPS receivers have almost completely replaced terrestrial instruments for large-scale base network surveys. For Planet-wide geodetic surveys, previously impossible, we can still mention Satellite Laser Ranging (SLR) and Lunar Laser Ranging (LLR) and Very Long Baseline Interferometry (VLBI) techniques. All these techniques also serve to monitor Earth rotation irregularities as well as plate tectonic motions.

Gravity is measured using gravimeters. Basically, there are two kinds of gravimeters. *Absolute* gravimeters, which nowadays can also be used in the field, are based directly on measuring the acceleration of free fall (for example, of a reflecting prism in a vacuum tube). They are used for establishing the vertical geospatial control. Most common *relative* gravimeters are spring based. They are used in gravity surveys over large areas for establishing the figure of the geoid over these areas. Most accurate relative gravimeters are *superconducting* gravimeters, and these are sensitive to one thousandth of one billionth of Earth surface gravity. Twenty-some superconducting gravimeters are used worldwide for studying Earth tides, rotation, interior, and ocean and atmospheric loading, as well as for verifying the Newtonian constant of gravitation.

In the future gravity, and altitude, will be measured by relativistic time dilation measured by strontium optical clocks.

Units and Measures on the Ellipsoid

Geographical latitude and longitude are stated in the units degree, minute of arc, and second of arc. They are *angles*, not metric measures, and describe the *direction* of the local normal to the reference ellipsoid of revolution. This is *approximately* the same as the direction of the plumbline, i.e., local gravity, which is also the normal to the geoid surface. For this reason, astronomical position determination – measuring the direction of the plumbline by astronomical means – works fairly well provided an ellipsoidal model of the figure of the Earth is used.

One geographical mile, defined as one minute of arc on the equator, equals 1,855.32571922 m. One nautical mile is one minute of astronomical latitude. The radius of curvature of the ellipsoid varies with latitude, being the longest at the pole and the shortest at the equator as is the nautical mile.

A metre was originally defined as the 10-millionth part of the length of a meridian (the target was not quite reached in actual implementation, so that is off by 200 ppm in the current definitions). This means that one kilometre is roughly equal to (1/40,000) * 360 * 60 meridional minutes of arc, which equals 0.54 nautical mile, though this is not exact because the two units are defined on different bases (the international nautical mile is defined as exactly 1,852 m, corresponding to a rounding of 1000/0.54 m to four digits).

Temporal Change

In geodesy, temporal change can be studied by a variety of techniques. Points on the Earth's surface change their location due to a variety of mechanisms:

- Continental plate motion, plate tectonics

- Episodic motion of tectonic origin, esp. close to fault lines

- Periodic effects due to Earth tides

- Postglacial land uplift due to isostatic adjustment

- Mass variations due to hydrological changes

- Various anthropogenic movements due to, for instance, petroleum or water extraction or reservoir construction.

The science of studying deformations and motions of the Earth's crust and the solid Earth as a whole is called geodynamics. Often, study of the Earth's irregular rotation is also included in its definition.

Techniques for studying geodynamic phenomena on the global scale include:

- satellite positioning by GPS and other such systems,

- Very Long Baseline Interferometry (VLBI)

- satellite and lunar laser ranging

- Regionally and locally, precise levelling,

- precise tacheometers,

- monitoring of gravity change,

- Interferometric synthetic aperture radar (InSAR) using satellite images, etc.

Geodetic Astronomy

Geodetic astronomy or astro-geodesy is the application of astronomical methods into networks and technical projects of geodesy.

The most important topics are:

- Establishment of geodetic datum systems (e.g. ED50) or at expeditions

- apparent places of stars, and their proper motions

- precise astronomical navigation

- astro-geodetic geoid determination

- modelling the rock densities of the topography and of geological layers in the subsurface

- Satellite geodesy using the background of stars

- Monitoring of the Earth rotation and polar wandering

- Contribution to the time system of physics and geosciences

Important measuring techniques are:

- Latitude and longitude determination by theodolites, tacheometers, astrolabes or zenith cameras

- time and star positions by observation of star transits, e.g. by meridian circles (visual, photographic or CCD)

- Azimuth measurements

- o   for the exact orientation of geodetic networks

- o   for mutual transformations between terrestrial and space methods

- o   for improved accuracy by means of "Laplace points" at special fixed points

- Vertical deflection measurements and their use

  - o   in geoid determination

  - o   in mathematical reduction of very precise networks

  - o   for geophysical and geological purposes

- Modern spatial methods

  - o   VLBI with radio sources (quasars)

  - o   Astrometry of stars by scanning satellites like Hipparcos or the future Gaia.

The accuracy of these methods depends on the instrument and its spectral wavelength, the measuring or scanning method, the time amount (versus economy), the atmospheric situation, the stability of the surface resp. the satellite, on mechanical and temperature effects to the instrument, on the experience and skill of the observer, and on the accuracy of the physical-mathematical models.

Therefore the accuracy reaches from 60" (navigation, ~1 mile) to 0,001" and better (a few cm; satellites, VLBI), e.g.

- angles (vertical deflections and azimuths) ±1" up to 0,1"

- geoid determination & height systems ca. 5 cm up to 0,2 cm

- astronomical lat/long and star positions ±1" up to 0,01"

- HIPPARCOS star positions ±0,001"

- VLBI quasar positions and Earth's rotation poles 0,001 to 0,0001" (cm-mm)

Satellite Geodesy

Wettzell Laser Ranging System, a satellite laser ranging station

Satellite geodesy is geodesy by means of artificial satellites — the measurement of the form and dimensions of Earth, the location of objects on its surface and the figure of the Earth's gravity field by means of artificial satellite techniques. It belongs to the broader field of space geodesy. Traditional astronomical geodesy is *not* commonly considered a part of satellite geodesy, although there is considerable overlap between the techniques.

The main goals of satellite geodesy are:

1. Determination of the figure of the Earth, positioning, and navigation (geometric satellite geodesy)

2. Determination of geoid, Earth's gravity field and its temporal variations (dynamical satellite geodesy)

3. Measurement of geodynamical phenomena, such as crustal dynamics and polar motion

Satellite geodetic data and methods can be applied to diverse fields such as navigation, hydrography, oceanography and geophysics. Satellite geodesy relies heavily on orbital mechanics.

History

First Steps (1957-1970)

Satellite geodesy began shortly after the launch of Sputnik in 1957. Observations of Explorer 1 and Sputnik 2 in 1958 allowed for an accurate determination of Earth's flattening. The 1960s saw the launch of the Doppler satellite Transit-1B and the balloon satellites Echo 1, Echo 2, and PAGEOS. The first dedicated geodetic satellite was ANNA-1B, a collaborative effort between NASA, the DoD, and other civilian agencies. ANNA-1B carried the first of the US Army's SECOR (Sequential Collation of Range) instruments. These missions led to the accurate determination of the leading spherical harmonic coefficients of the geopotential, the general shape of the geoid, and linked the world's geodetic datums.

Soviet military satellites undertook geodesic missions to assist in ICBM targeting in the late 1960s and early 1970s.

Toward the World Geodetic System (1970-1990)

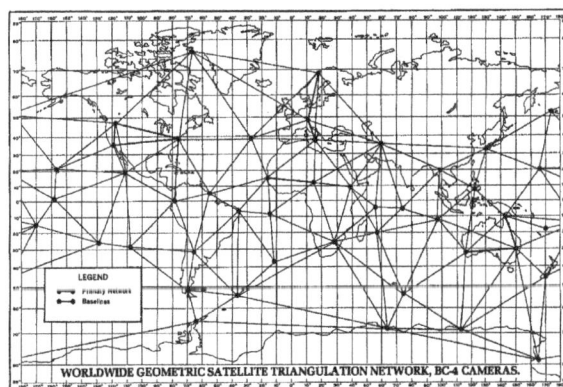

Worldwide BC-4 camera geometric satellite triangulation network

The Transit satellite system was used extensively for Doppler surveying, navigation, and positioning. Observations of satellites in the 1970s by worldwide triangulation networks allowed for the establishment of the World Geodetic System. The development of GPS by the United States in the 1980s allowed for precise navigation and positioning and soon became a standard tool in surveying. In the 1980s and 1990s satellite geodesy began to be used for monitoring of geodynamic phenomena, such as crustal motion, Earth rotation, and polar motion.

Modern Era (1990-present)

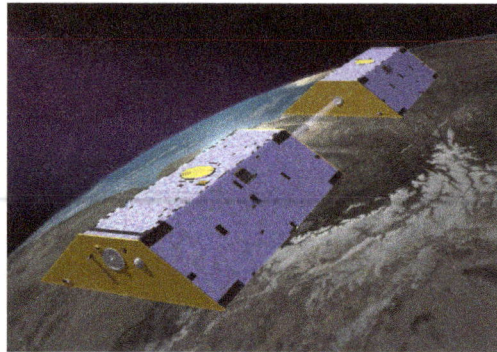

Artist's conception of GRACE

The 1990s were focused on the development of permanent geodetic networks and reference frames. Dedicated satellites were launched to measure Earth's gravity field in the 2000s, such as CHAMP, GRACE, and GOCE.

Satellite Geodetic Measurement Techniques

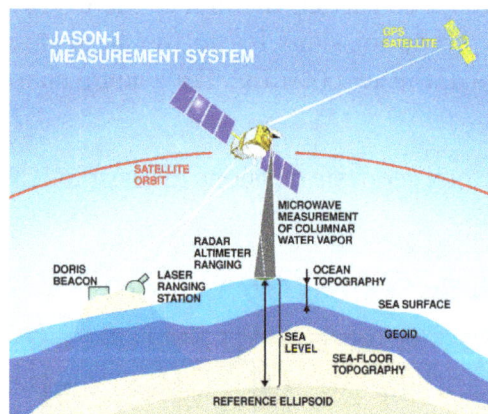

The Jason-1 measurement system combines major geodetic measurement techniques, including DORIS, SLR, GPS, and altimetry.

Techniques of satellite geodesy may be classified by instrument platform: A satellite may

1. be observed with ground-based instruments (*Earth-to-space-methods*),

2. carry an instrument or sensor as part of its payload to observe the Earth (*space-to-Earth methods*),

3. or use its instruments to track or be tracked by another satellite (*space-to-space methods*).

Earth-to-Space Methods

Geodetic use of GPS/GNSS

Global navigation satellite systems are dedicated radio positioning services, which can locate a receiver to within a few meters. The most prominent system, GPS, consists of a constellation of 31 satellites (as of December 2013) in high, 12-hour circular orbits, distributed in six planes with 55° inclinations. The principle of location is based on trilateration. Each satellite transmits a precise ephemeris with information on its own position and a message containing the exact time of transmission. The receiver compares this time of transmission with its own clock at the time of reception and multiplies the difference by the speed of light to obtain a "pseudorange." Four pseudoranges are needed to obtain the precise time and the receiver's position within a few meters. More sophisticated methods, such as real-time kinematic (RTK) can yield positions to within a few millimeters.

In geodesy, GNSS is used as an economical tool for surveying and time transfer. It is also used for monitoring Earth's rotation, polar motion, and crustal dynamics. The presence of the GPS signal in space also makes it suitable for orbit determination and satellite-to-satellite tracking. *Examples: GPS, GLONASS, Galileo.*

Laser Ranging

In satellite laser ranging (SLR) a global network of observation stations measure the round trip time of flight of ultrashort pulses of light to satellites equipped with retroreflectors. This provides instantaneous range measurements of millimeter level precision which can be accumulated to provide accurate orbit parameters, gravity field parameters (from the orbit perturbations), Earth rotation parameters, tidal Earth's deformations, coordinates and velocities of SLR stations, and other substantial geodetic data. Satellite laser ranging is a proven geodetic technique with significant potential for important contributions to scientific studies of the Earth/Atmosphere/Oceans system. It is the most accurate technique currently available to determine the geocentric position of an Earth satellite, allowing for the precise calibration of radar altimeters and separation of long-term instrumentation drift from secular changes in ocean surface topography. Satellite laser ranging contributes to the definition of the international terrestrial reference frames by providing the information about the scale and the origin of the reference frame, the so-called geocenter coordinates. *Example: LAGEOS.*

Doppler Techniques

Doppler positioning involves recording the Doppler shift of a radio signal of stable frequency emitted from a satellite as the satellite approaches and recedes from the observer. The observed frequency depends on the radial velocity of the satellite relative to the observer, which is constrained by orbital mechanics. If the observer knows the orbit of the satellite, then the recording the Doppler profile determines the observer's position. Conversely, if the observer's position is precisely known, then the orbit of the satellite can be determined and used to study the Earth's gravity. In DORIS, the ground station emits the signal and the satellite receives. *Examples: Transit, DORIS.*

Optical Tracking

In optical tracking, the satellite can be used as a very high target for triangulation and can be used to ascertain the geometric relationship between multiple observing stations. Optical tracking with the BC-4, PC-1000, MOTS, or Baker Nunn cameras consisted of photographic observations of a satellite, or flashing light on the satellite, against a background of stars. The stars, whose positions were accurately determined, provided a framework on the photographic plate or film for a determination of precise directions from camera station to satellite. Geodetic positioning work with cameras was usually performed with one camera observing simultaneously with one or more other cameras. Camera systems are weather dependent and that is one major reason why they fell out of use by the 1980s. *Examples: PAGEOS, Project Echo.*

Space-to-Earth Methods

Radar Altimetry

A radar altimeter uses the round-trip flight-time of a microwave pulse between the satellite and the Earth's surface to determine the distance between the spacecraft and the surface. From this distance or height, the local surface effects such as tides, winds and currents are removed to obtain the satellite height above the geoid. With a precise ephemeris available for the satellite, the geocentric position and ellipsoidal height of the satellite are available for any given observation time. It is then possible to compute the geoid height by subtracting the measured altitude from the ellipsoidal height. This allows direct measurement of the geoid, since the ocean surface closely follows the geoid. The difference between the ocean surface and the actual geoid gives ocean surface topography. *Examples: Seasat, Geosat, TOPEX/Poseidon, ERS-1, ERS-2, Jason-1, Jason-2, Envisat.*

Laser Altimetry

A laser altimeter uses the round-trip flight-time of a beam of light at optical or infrared wavelengths to determine the spacecraft's altitude. *Example: ICESat.*

Interferometric Synthetic Aperture Radar (InSAR)

Interferometric synthetic aperture radar (InSAR) is a radar technique used in geodesy and remote sensing. This geodetic method uses two or more synthetic aperture radar (SAR) images to generate maps of surface deformation or digital elevation, using differences in the phase of the waves returning to the satellite. The technique can potentially measure centimetre-scale changes in deformation over timespans of days to years. It has applications for geophysical monitoring of natural hazards, for example earthquakes, volcanoes and landslides, and also in structural engineering, in particular monitoring of subsidence and structural stability. *Example: Seasat, TerraSAR-X.*

Gravity Gradiometry

A gravity gradiometer can independently determine the components of the gravity vector on a real-time basis. A gravity gradient is simply the spatial derivative of the gravity vector. The gradient can be thought of as the rate of change of a component of the gravity vector as measured over a small distance. Hence, the gradient can be measured by determining the difference in gravity at two close but distinct points. This principle is embodied in several recent moving-base instruments. The gravity gradient at a point is a tensor, since it is the derivative of each component of the gravity vector taken in each sensitive axis. Thus, the value of any component of the gravity vector

can be known all along the path of the vehicle if gravity gradiometers are included in the system and their outputs are integrated by the system computer. An accurate gravity model will be computed in real-time and a continuous map of normal gravity, elevation, and anomalous gravity will be available. *Example: GOCE.*

Space-to-Space Methods

Satellite-to-Satellite Tracking

This technique uses satellites to track other satellites. There are a number of variations which may be used for specific purposes such as gravity field investigations and orbit improvement.

- A high altitude satellite may act as a relay from ground tracking stations to a low altitude satellite. In this way, low altitude satellites may be observed when they are not accessible to ground stations. In this type of tracking, a signal generated by a tracking station is received by the relay satellite and then retransmitted to a lower altitude satellite. This signal is then returned to the ground station by the same path.

- Two low altitude satellites can track one another observing mutual orbital variations caused by gravity field irregularities. A prime example of this is GRACE.

- Several high altitude satellites with accurately known orbits, such as GPS satellites, may be used to fix the position of a low altitude satellite.

These examples present a few of the possibilities for the application of satellite-to-satellite tracking. Satellite-to-satellite tracking data was first collected and analyzed in a high-low configuration between ATS-6 and GEOS-3. The data was studied to evaluate its potential for both orbit and gravitational model refinement.

Physical Geodesy

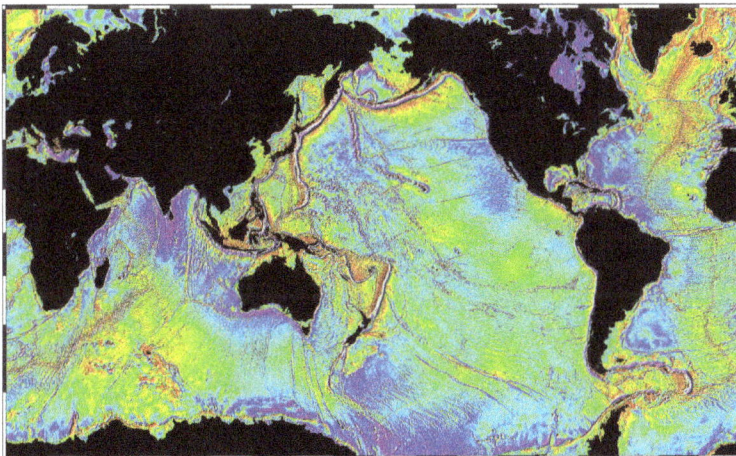

Ocean basins mapped with satellite altimetry. Seafloor features larger than 10 km are detected by resulting gravitational distortion of sea surface. (1995, NOAA)

Physical geodesy is the study of the physical properties of the gravity field of the Earth, the geopotential, with a view to their application in geodesy.

Measurement Procedure

Traditional geodetic instruments such as theodolites rely on the gravity field for orienting their vertical axis along the local plumb line or local vertical direction with the aid of a spirit level. After that, vertical angles (zenith angles or, alternatively, elevation angles) are obtained with respect to this local vertical, and horizontal angles in the plane of the local horizon, perpendicular to the vertical.

Levelling instruments again are used to obtain geopotential differences between points on the Earth's surface. These can then be expressed as "height" differences by conversion to metric units.

The Geopotential

The Earth's gravity field can be described by a potential as follows:

$$\mathbf{g} = \nabla W = \text{grad } W = \frac{\partial W}{\partial X}\mathbf{i} + \frac{\partial W}{\partial Y}\mathbf{j} + \frac{\partial W}{\partial Z}\mathbf{k}$$

which expresses the gravitational acceleration vector as the gradient of W, the potential of gravity. The vector triad $\{\mathbf{i}, \mathbf{j}, \mathbf{k}\}$ is the orthonormal set of base vectors in space, pointing along the X, Y, Z coordinate axes.

Note that both gravity and its potential contain a contribution from the centrifugal pseudo-force due to the Earth's rotation. We can write

$$W = V + \Phi$$

where V is the potential of the *gravitational* field, W that of the *gravity* field, and Φ that of the centrifugal force field.

The centrifugal force -- per unit of mass, i.e., acceleration -- is given by

$$\mathbf{g}_c = \omega^2 \mathbf{p},$$

where

$$\mathbf{p} = X\mathbf{i} + Y\mathbf{j} + 0 \cdot \mathbf{k}$$

is the vector pointing to the point considered straight from the Earth's rotational axis. It can be shown that this pseudo-force field, in a reference frame co-rotating with the Earth, has a potential associated with it that looks like this:

$$\Phi = \frac{1}{2}\omega^2(X^2 + Y^2).$$

This can be verified by taking the gradient (∇) operator of this expression.

Here, X, Y and Z are geocentric coordinates.

Units of Gravity and Geopotential

Gravity is commonly measured in units of m·s⁻² (metres per second squared). This also can be expressed (multiplying by the gravitational constant **G** in order to change units) as newtons per kilogram of attracted mass.

Potential is expressed as gravity times distance, m²·s⁻². Travelling one metre in the direction of a gravity vector of strength 1 m·s⁻² will increase your potential by 1 m²·s⁻². Again employing G as a multipier, the units can be changed to joules per kilogram of attracted mass.

A more convenient unit is the GPU, or geopotential unit: it equals 10 m²·s⁻². This means that travelling one metre in the vertical direction, i.e., the direction of the 9.8 m·s⁻² ambient gravity, will *approximately* change your potential by 1 GPU. Which again means that the difference in geopotential, in GPU, of a point with that of sea level can be used as a rough measure of height "above sea level" in metres.

The Normal Potential

To a rough approximation, the Earth is a sphere, or to a much better approximation, an ellipsoid. We can similarly approximate the gravity field of the Earth by a spherically symmetric field:

$$W \approx \frac{GM}{R}$$

of which the *equipotential surfaces*—the surfaces of constant potential value—are concentric spheres.

It is more accurate to approximate the geopotential by a field that has *the Earth reference ellipsoid* as one of its equipotential surfaces, however. The most recent Earth reference ellipsoid is GRS80, or Geodetic Reference System 1980, which the Global Positioning system uses as its reference. Its geometric parameters are: semi-major axis a = 6378137.0 m, and flattening f = 1/298.257222101.

A geopotential field U is constructed, being the sum of a gravitational potential Ψ and the known centrifugal potential Φ, that *has the GRS80 reference ellipsoid as one of its equipotential surfaces*. If we also require that the enclosed mass is equal to the known mass of the Earth (including atmosphere) GM = 3986005 × 10⁸ m³·s⁻², we obtain for the *potential at the reference ellipsoid:*

$$U_0 = 62636860.850 \, \text{m}^2\text{s}^{-2}$$

Obviously, this value depends on the assumption that the potential goes asymptotically to zero at infinity ($R \to \infty$), as is common in physics. For practical purposes it makes more sense to choose the zero point of normal gravity to be that of the reference ellipsoid, and refer the potentials of other points to this.

Disturbing Potential and Geoid

Once a clean, smooth geopotential field U has been constructed matching the known GRS80 reference ellipsoid with an equipotential surface (we call such a field a *normal potential*) we can subtract it from the true (measured) potential W of the real Earth. The result is defined as **T**, the *disturbing potential*:

$$T = W - U$$

The disturbing potential T is numerically a great deal smaller than U or W, and captures the detailed, complex variations of the true gravity field of the actually existing Earth from point-to-point, as distinguished from the overall global trend captured by the smooth mathematical ellipsoid of the normal potential.

Due to the irregularity of the Earth's true gravity field, the equilibrium figure of sea water, or the geoid, will also be of irregular form. In some places, like west of Ireland, the geoid—mathematical mean sea level—sticks out as much as 100 m above the regular, rotationally symmetric reference ellipsoid of GRS80; in other places, like close to Ceylon, it dives under the ellipsoid by nearly the same amount. The separation between these two surfaces is called the undulation of the geoid, symbol N , and is closely related to the disturbing potential.

According to the famous Bruns formula, we have

$$N = T / \gamma,$$

where γ is the force of gravity computed from the normal field potential U .

In 1849, the mathematician George Gabriel Stokes published the following formula named after him:

$$N = \frac{R}{4\pi\gamma_0} \iint_\sigma \Delta g S(\psi) d\sigma.$$

In this formula, Δg stands for *gravity anomalies*, differences between true and normal (reference) gravity, and S is the *Stokes function*, a kernel function derived by Stokes in closed analytical form. (Note that determining N anywhere on Earth by this formula requires Δg to be known *everywhere on Earth*. Welcome to the role of international co-operation in physical geodesy.)

The geoid, or mathematical mean sea surface, is defined not only on the seas, but also under land; it is the equilibrium water surface that would result, would sea water be allowed to move freely (e.g., through tunnels) under the land. Technically, an *equipotential surface* of the true geopotential, chosen to coincide (on average) with mean sea level.

As mean sea level is physically realized by tide gauge bench marks on the coasts of different countries and continents, a number of slightly incompatible "near-geoids" will result, with differences of several decimetres to over one metre between them, due to the dynamic sea surface topography. These are referred to as *vertical* or *height datums*.

For every point on Earth, the local direction of gravity or vertical direction, materialized with the plumb line, is *perpendicular* to the geoid. On this is based a method, *astrogeodetic levelling*, for deriving the local figure of the geoid by measuring *deflections of the vertical* by astronomical means over an area.

Gravity Anomalies

Above we already made use of *gravity anomalies* Δg. These are computed as the differences between true (observed) gravity $g = \| \vec{g} \|$, and calculated (normal) gravity $\gamma = \| \vec{\gamma} \| = \| \nabla U \|$. (This is an oversimplification; in practice the location in space at which γ is evaluated will differ slightly from that where g has been measured.) We thus get

$$\Delta g = g - \gamma.$$

These anomalies are called free-air anomalies, and are the ones to be used in the above Stokes equation.

In geophysics, these anomalies are often further reduced by removing from them the *attraction of the topography*, which for a flat, horizontal plate (Bouguer plate) of thickness H is given by

$$a_B = 2\pi G \rho H,$$

The Bouguer reduction to be applied as follows:

$$\Delta g_B = \Delta g_{FA} - a_B,$$

so-called Bouguer anomalies. Here, Δg_{FA} is our earlier Δg, the free-air anomaly.

In case the terrain is not a flat plate (the usual case!) we use for H the local terrain height value but apply a further correction called the terrain correction (*TC*).

History of Geodesy

Geodesy also named geodetics, is the scientific discipline that deals with the measurement and representation of the Earth. The history of geodesy began in antiquity and blossomed during the Age of Enlightenment.

Early ideas about the figure of the Earth held the Earth to be flat, and the heavens a physical dome spanning over it. Two early arguments for a spherical Earth were that lunar eclipses were seen as circular shadows which could only be caused by a spherical Earth, and that Polaris is seen lower in the sky as one travels South.

Hellenic World

The early Greeks, in their speculation and theorizing, ranged from the flat disc advocated by Homer to the spherical body postulated by Pythagoras. Pythagoras's idea was supported later by Aristotle. Pythagoras was a mathematician and to him the most perfect figure was a sphere. He reasoned that the gods would create a perfect figure and therefore the Earth was created to be spherical in shape. Anaximenes, an early Greek philosopher, believed strongly that the Earth was rectangular in shape.

Since the spherical shape was the most widely supported during the Greek Era, efforts to determine its size followed. Plato determined the circumference of the Earth (which is slightly over 40,000 km) to be 400,000 stadia (between 62,800 and 74,000 km or 46,250 and 39,250 mi) while Archimedes estimated 300,000 stadia (48,300 km or 30,000 mi), using the Hellenic stadion which scholars generally take to be 185 meters or $\frac{1}{10}$ of a geographical mile. Plato's figure was a guess and Archimedes' a more conservative approximation.

Hellenistic World

In Egypt, a Greek scholar and philosopher, Eratosthenes (276 BC – 195 BC), is said to have made more explicit measurements. He had heard that on the longest day of the summer solstice, the midday sun shone to the bottom of a well in the town of Syene (Aswan). At the same time, he observed the sun was not directly overhead at Alexandria; instead, it cast a shadow with the vertical equal to 1/50th of a circle (7° 12'). To these observations, Eratosthenes applied certain "known" facts (1) that on the day of the summer solstice, the midday sun was directly over the Tropic of Cancer; (2) Syene was on this tropic; (3) Alexandria and Syene lay on a direct north-south line; (4) The sun was a relatively long way away (astronomical unit). Legend has it that he had someone walk from Alexandria to Syene to measure the distance, which came out to be equal to 5000 stadia or (at the usual Hellenic 185 m per stadion) about 925 km.

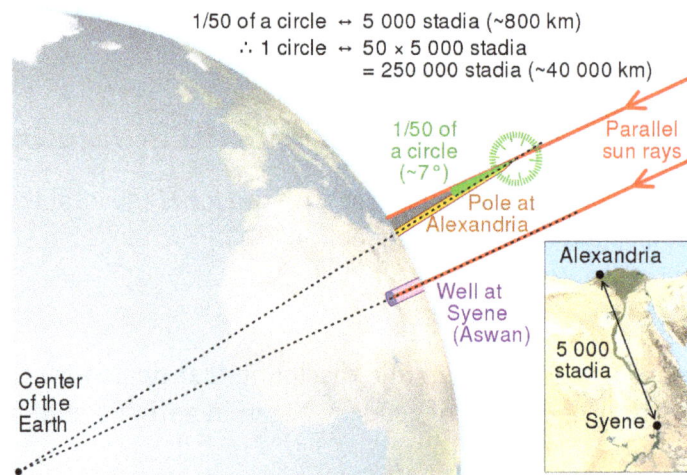

Eratosthenes' method for determining the size of the Earth

From these observations, measurements, and/or "known" facts, Eratosthenes concluded that since the angular deviation of the sun from the vertical direction at Alexandria was also the angle of the subtended arc, the linear distance between Alexandria and Syene was 1/50 of the circumference of the Earth which thus must be 50×5000 = 250,000 stadia or probably 25,000 geographical miles. The circumference of the Earth is 24,902 mi (40,075.16 km). Over the poles it is more precisely 40,008 km or 24,860 mi. The actual unit of measure used by Eratosthenes was the stadion. No one knows for sure what his stadion equals in modern units, but some say that it was the Hellenic 185 m stadion.

Had the experiment been carried out as described, it would not be remarkable if it agreed with actuality. What is remarkable is that the result was probably only about 0.4% too high. His measurements were subject to several inaccuracies: (1) though at the summer solstice the noon sun is overhead at the Tropic of Cancer, Syene was not exactly on the tropic (which was at 23° 43' latitude in that day) but about 22 geographical miles to the north; (2) the difference of latitude between Alexandria (31.2 degrees north latitude) and Syene (24.1 degrees) is really 7.1 degrees rather than the perhaps rounded (1/50 of a circle) value of 7° 12' that Eratosthenes used; (3) the actual solstice zenith distance of the noon sun at Alexandria was 31° 12' – 23° 43' = 7° 29' or about 1/48 of a circle not 1/50 = 7° 12', an error closely consistent with use of a vertical gnomon which fixes not the sun's center but the solar upper limb 16' higher; (4) the most importantly flawed element, whether he measured or adopted it, was the latitudinal distance from Alexandria to Syene (or the true Tropic somewhat further south) which he appears to have overestimated by a factor that relates to most of the error in his resulting circumference of the earth.

A parallel later ancient measurement of the size of the Earth was made by another Greek scholar, Posidonius. He is said to have noted that the star Canopus was hidden from view in most parts of Greece but that it just grazed the horizon at Rhodes. Posidonius is supposed to have measured the elevation of Canopus at Alexandria and determined that the angle was 1/48 of circle. He assumed the distance from Alexandria to Rhodes to be 5000 stadia, and so he computed the Earth's circumference in stadia as 48 times 5000 = 240,000. Some scholars see these results as luckily semi-accurate due to cancellation of errors. But since the Canopus observations are both mistaken by over a degree, the "experiment" may be not much more than a recycling of Eratosthenes's numbers, while altering 1/50 to the correct 1/48 of a circle. Later, either he or a follower appears to have altered the base distance to agree with Eratosthenes's Alexandria-to-Rhodes figure of 3750 stadia since Posidonius's final circumference was 180,000 stadia, which equals 48×3750 stadia. The 180,000 stadia circumference of Posidonius is suspiciously close to that which results from another method of measuring the earth, by timing ocean sunsets from different heights, a method which is inaccurate due to horizontal atmospheric refraction.

The abovementioned larger and smaller sizes of the Earth were those used by Claudius Ptolemy at different times, 252,000 stadia in his *Almagest* and 180,000 stadia in his later *Geography*. His midcareer conversion resulted in the latter work's systematic exaggeration of degree longitudes in the Mediterranean by a factor close to the ratio of the two seriously differing sizes discussed here, which indicates that the conventional size of the earth was what changed, not the stadion.

Ancient India

The Indian mathematician Aryabhata (AD 476–550) was a pioneer of mathematical astronomy. He describes the earth as being spherical and that it rotates on its axis, among other things in his work Āryabhaṭīya. Aryabhatiya is divided into four sections. Gitika, Ganitha (mathematics), Kalakriya (reckoning of time) and Gola (celestial sphere). The discovery that the earth rotates on its own axis from west to east is described in Aryabhatiya (Gitika 3,6; Kalakriya 5; Gola 9,10;). For example, he explained the apparent motion of heavenly bodies is only an illusion (Gola 9), with the following simile;

> Just as a passenger in a boat moving downstream sees the stationary (trees on the river banks) as traversing upstream, so does an observer on earth see the fixed stars as moving towards the west at exactly the same speed (at which the earth moves from west to east.)

Aryabhatiya also estimates the circumference of Earth, with an error of 1%, which is remarkable. Aryabhata gives the radii of the orbits of the planets in terms of the Earth-Sun distance as essentially their periods of rotation around the Sun. He also gave the correct explanation of lunar and solar eclipses and that the Moon shines by reflecting sunlight.

Islamic World

The Muslim scholars, who held to the spherical Earth theory, used it to calculate the distance and direction from any given point on the earth to Mecca. This determined the Qibla, or Muslim direction of prayer. Muslim mathematicians developed spherical trigonometry which was used in these calculations.

Around AD 830 Caliph al-Ma'mun commissioned a group of astronomers led by Al-Khwarizmi to measure the distance from Tadmur (Palmyra) to Raqqa, in modern Syria. They found the cities to be separated by one degree of latitude and the distance between them to be $66\frac{2}{3}$ miles and thus calculated the Earth's circumference to be 24,000 miles. Another estimate given was $56\frac{2}{3}$ Arabic miles per degree, which corresponds to 111.8 km per degree and a circumference of 40,248 km, very close to the currently modern values of 111.3 km per degree and 40,068 km circumference, respectively.

Muslim astronomers and geographers were aware of magnetic declination by the 15th century, when the Egyptian astronomer 'Abd al-'Aziz al-Wafa'i (d. 1469/1471) measured it as 7 degrees from Cairo.

Al-Biruni

Of the medieval Persian Abu Rayhan al-Biruni (973–1048) it is said:

"Important contributions to geodesy and geography were also made by Biruni. He introduced techniques to measure the earth and distances on it using triangulation. He found the radius of the earth to be 6339.6 km, a value not obtained in the West until the 16th century. His *Masudic canon* contains a table giving the coordinates of six hundred places, almost all of which he had direct knowledge."

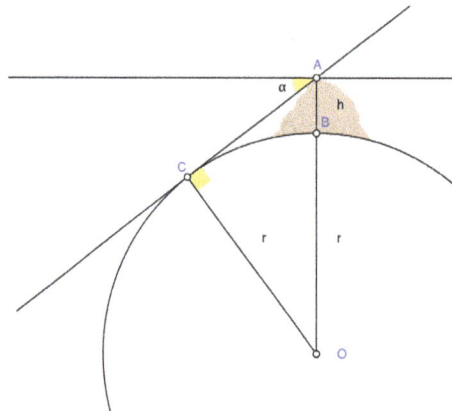

Abu Rayhan al-Biruni accurately determined the Earth radius by formulating a trigonometric equation relating the dip angle (between the true horizon and astronomical horizon) observed from the top of a mountain to the height of that mountain.

At the age of 17, Al-Biruni calculated the latitude of Kath, Khwarazm, using the maximum altitude of the Sun. Al-Biruni also solved a complex geodesic equation in order to accurately compute the Earth's circumference, which were close to modern values of the Earth's circumference. His estimate of 6,339.9 km for the Earth radius was only 16.8 km less than the modern value of 6,356.7523142 km (WGS84 polar radius "b"). In contrast to his predecessors who measured the

Earth's circumference by sighting the Sun simultaneously from two different locations, Al-Biruni developed a new method of using trigonometric calculations based on the angle between a plain and mountain top which yielded more accurate measurements of the Earth's circumference and made it possible for it to be measured by a single person from a single location. Al-Biruni's method's motivation was to avoid "walking across hot, dusty deserts" and the idea came to him when he was on top of a tall mountain in India (present day Pind Dadan Khan, Pakistan). From the top of the mountain, he sighted the dip angle which, along with the mountain's height (which he calculated beforehand), he applied to the law of sines formula. This was the earliest known use of dip angle and the earliest practical use of the law of sines. He also made use of algebra to formulate trigonometric equations and used the astrolabe to measure angles. His method can be summarized as follows:

He first calculated the height of the mountain by going to two points at sea level with a known distance apart and then measuring the angle between the plain and the top of the mountain for both points. He made both the measurements using an astrolabe. He then used the following trigonometric formula relating the distance (d) between both points with the tangents of their angles (θ) to determine the height (h) of the mountain:

$$h = \frac{d \tan \theta_1 \tan \theta_2}{\tan \theta_2 - \tan \theta_1}$$

He then stood at the highest point of the mountain, where he measured the dip angle using an astrolabe. He applied the values he obtained for the dip angle and the mountain's height to the following trigonometric formula in order to calculate the Earth's radius:

$$R = \frac{h \cos \theta}{1 - \cos \theta}$$

where

- R = Earth radius
- h = height of mountain
- θ = dip angle

Al-Biruni had also, by the age of 22, written a study of map projections, *Cartography*, which included a method for projecting a hemisphere on a plane. Around 1025, Al-Biruni was the first to describe a polar equi-azimuthal equidistant projection of the celestial sphere. He was also regarded as the most skilled when it came to mapping cities and measuring the distances between them, which he did for many cities in the Middle East and western Indian subcontinent. He often combined astronomical readings and mathematical equations, in order to develop methods of pin-pointing locations by recording degrees of latitude and longitude. He also developed similar techniques when it came to measuring the heights of mountains, depths of valleys, and expanse of the horizon, in *The Chronology of the Ancient Nations*. He also discussed human geography and the planetary habitability of the Earth. He hypothesized that roughly a quarter of the Earth's surface is habitable by humans, and also argued that the shores of Asia and Europe were "separated by a vast sea, too dark and dense to navigate and too risky to try".

Medieval Europe

Revising the figures attributed to Posidonius, another Greek philosopher determined 18,000 miles as the Earth's circumference. This last figure was promulgated by Ptolemy through his world maps. The maps of Ptolemy strongly influenced the cartographers of the Middle Ages. It is probable that Christopher Columbus, using such maps, was led to believe that Asia was only 3 or 4 thousand miles west of Europe.

Ptolemy's view was not universal, however, and chapter 20 of *Mandeville's Travels* (c. 1357) supports Eratosthenes' calculation.

It was not until the 16th century that his concept of the Earth's size was revised. During that period the Flemish cartographer, Mercator, made successive reductions in the size of the Mediterranean Sea and all of Europe which had the effect of increasing the size of the earth.

Early Modern Period

The invention of the telescope and the theodolite and the development of logarithm tables allowed exact triangulation and grade measurement.

Europe

In 1505 the cosmographer and explorer Duarte Pacheco Pereira calculated the value of the degree of the meridian arc with a margin of error of only 4%, when the current error at the time varied between 7 and 15%.

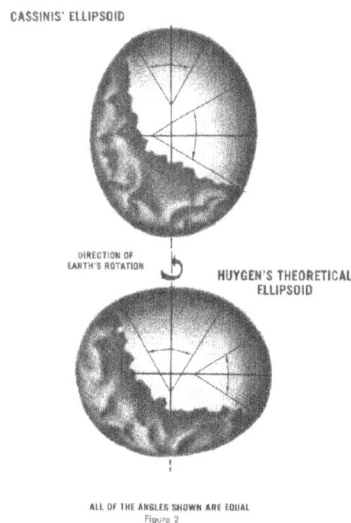

Cassini's ellipsoid; Huygens' theoretical ellipsoid

Jean Picard performed the first modern meridian arc measurement in 1669–1670. He measured a baseline using wooden rods, a telescope (for his angular measurements), and logarithms (for computation). Jacques Cassini later continued Picard's arc northward to Dunkirk and southward to the Spanish border. Cassini divided the measured arc into two parts, one northward from Paris, another southward. When he computed the length of a degree from both chains, he found that the length of one degree of latitude in the northern part of the chain was shorter than that in the southern part.

This result, if correct, meant that the earth was not a sphere, but a prolate spheroid (taller than wide). However, this contradicted computations by Isaac Newton and Christiaan Huygens. Newton's theory of gravitation predicted the Earth to be an oblate spheroid (wider than tall), with a flattening of 1:230.

The issue could be settled by measuring, for a number of points on earth, the relationship between their distance (in north-south direction) and the angles between their zeniths. On an oblate Earth, the meridional distance corresponding to one degree of longitude will grow toward the poles, as can be demonstrated mathematically.

The French Academy of Sciences dispatched two expeditions. One expedition (1736–37) under Pierre Louis Maupertuis was sent to Torne Valley (near the Earth's northern pole). The second mission (1735–44) under Pierre Bouguer was sent to what is modern-day Ecuador, near the equator. Their measurements demonstrated an oblate Earth, with a flattening of 1:210. This approximation to the true shape of the Earth became the new reference ellipsoid.

Asia and Americas

In South America Bouguer noticed, as did George Everest in the 19th century Great Trigonometric Survey of India, that the astronomical vertical tended to be pulled in the direction of large mountain ranges, due to the gravitational attraction of these huge piles of rock. As this vertical is everywhere perpendicular to the idealized surface of mean sea level, or the geoid, this means that the figure of the Earth is even more irregular than an ellipsoid of revolution. Thus the study of the "undulation of the geoid" became the next great undertaking in the science of studying the figure of the Earth.

19th Century

Archive with lithography plates for maps of Bavaria in the *Landesamt für Vermessung und Geoinformation* in Munich.

In the late 19th century the Zentralbüro für die Internationale Erdmessung (Central Bureau for International Geodesy) was established by Austria-Hungary and Germany. One of its most important goals was the derivation of an international ellipsoid and a gravity formula which should be optimal not only for Europe but also for the whole world. The Zentralbüro was an early predecessor of the International Association of Geodesy (IAG) and the International Union of Geodesy and Geophysics (IUGG) which was founded in 1919.

Negative lithography stone and positive print of a historic map of Munich Louis Puissant, *Traité de géodésie*, 1842.

Most of the relevant theories were derived by the German geodesist Friedrich Robert Helmert in his famous books *Die mathematischen und physikalischen Theorieen der höheren Geodäsie*, Einleitung und 1. Teil (1880) and 2. Teil (1884); English translation: Mathematical and Physical Theories of Higher Geodesy, Vol. 1 and Vol. 2. Helmert also derived the first global ellipsoid in 1906 with an accuracy of 100 meters (0.002 percent of the Earth's radii). The US geodesist Hayford derived a global ellipsoid in ~1910, based on intercontinental isostasy and an accuracy of 200 m. It was adopted by the IUGG as "international ellipsoid 1924".

References

- Sosnica, Krzysztof (2014). Determination of Precise Satellite Orbits and Geodetic Parameters using Satellite Laser Ranging. Bern: Astronomical Institute, University of Bern, Switzerland. p. 6. ISBN 8393889804

- Massonnet, D.; Feigl, K. L. (1998), "Radar interferometry and its application to changes in the earth's surface", Rev. Geophys., 36 (4), pp. 441–500, Bibcode:1998RvGeo..36..441M, doi:10.1029/97RG03139

- Hanssen, Ramon F. (2001), Radar Interferometry: Data Interpretation and Error Analysis, Kluwer Academic, ISBN 9780792369455

- Burgmann, R.; Rosen, P.A.; Fielding, E.J. (2000), "Synthetic aperture radar interferometry to measure Earth's surface topography and its deformation", Annual Review of Earth and Planetary Sciences, 28, pp. 169–209, Bibcode:2000AREPS..28..169B, doi:10.1146/annurev.earth.28.1.169

- M .R. Hoare: Quest for the true figure of the Earth: ideas and expeditions in four centuries of geodesy. Burlington, VT: Ashgate, 2004 ISBN 0-7546-5020-0

2

Understanding Gravity of Earth

Gravity of the Earth can be denoted by the symbol g. The force of the gravity of Earth depends on it location. Some of the topics discussed related to the gravity of Earth are theoretical gravity, gravimetry, gravitation of the moon, gravimeter and gravity gradiometry. This chapter elucidates the crucial theories and principles of the Earth's gravity.

Gravity of Earth

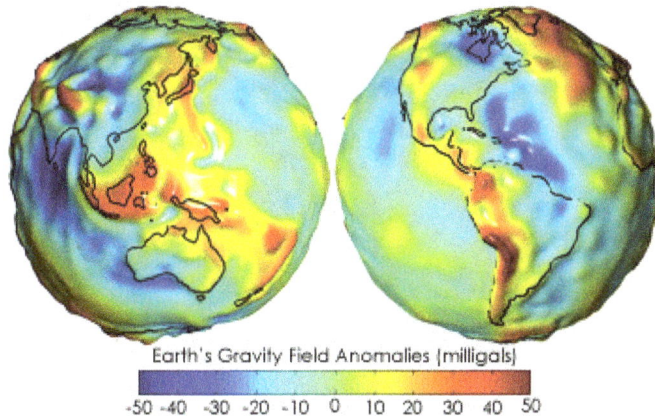

Earth's gravity measured by NASA GRACE mission, showing deviations from the theoretical gravity of an idealized smooth Earth, the so-called earth ellipsoid. Red shows the areas where gravity is stronger than the smooth, standard value, and blue reveals areas where gravity is weaker.

The gravity of Earth, which is denoted by g, refers to the acceleration that is imparted to objects due to the distribution of mass within the Earth. In SI units this acceleration is measured in metres per second squared (in symbols, m/s² or m·s⁻²) or equivalently in newtons per kilogram (N/kg or N·kg⁻¹). Near the Earth's surface, gravitational acceleration is approximately 9.8 m/s², which means that, ignoring the effects of air resistance, the speed of an object falling freely will increase by about 9.8 metres (32 ft) per second every second. This quantity is sometimes referred to informally as *little g* (in contrast, the gravitational constant G is referred to as *big G*).

The precise strength of Earth's gravity varies depending on location. The nominal "average" value at the Earth's surface, known as standard gravity is, by definition, 9.80665 m/s² (about 32.1740 ft/s²). This quantity is denoted variously as g_n, g_e (though this sometimes means the normal equatorial value on Earth, 9.78033 m/s²), g_0, gee, or simply g (which is also used for the variable local value). The weight of an object on the Earth's surface is the downwards force on that object, given by Newton's second law of motion, or $F = ma$ (*force = mass × acceleration*). Gravitational acceleration contributes to the total acceleration, but other factors, such as the rotation of the Earth, also contribute, and, therefore, affect the weight of the object.

Variation in Gravity and Apparent Gravity

A perfect sphere of uniform mass density, or whose density varies solely with distance from the centre (spherical symmetry), would produce a gravitational field of uniform magnitude at all points on its surface, always pointing directly towards the sphere's centre. The Earth is not spherically symmetric, but is slightly flatter at the poles while bulging at the Equator: an oblate spheroid. There are consequently slight deviations in both the magnitude and direction of gravity across its surface. The net force (or corresponding net acceleration) as measured by a scale and plumb bob is called "effective gravity" or "apparent gravity". Effective gravity includes other factors that affect the net force. These factors vary and include things such as centrifugal force at the surface from the Earth's rotation and the gravitational pull of the Moon and Sun.

Effective gravity on the Earth's surface varies by around 0.7%, from 9.7639 m/s² on the Nevado Huascarán mountain in Peru to 9.8337 m/s² at the surface of the Arctic Ocean. In large cities, it ranges from 9.7760 in Kuala Lumpur, Mexico City, and Singapore to 9.825 in Oslo and Helsinki.

Latitude

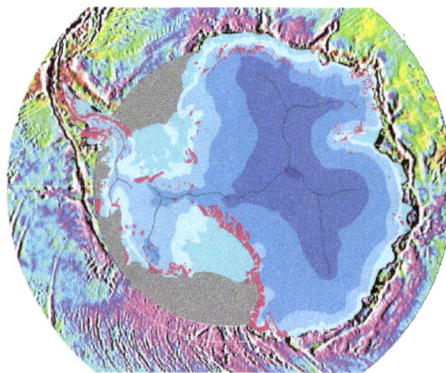

The differences of Earth's gravity around the Antarctic continent.

The surface of the Earth is rotating, so it is not an inertial frame of reference. At latitudes nearer the Equator, the outward centrifugal force produced by Earth's rotation is larger than at polar latitudes. This counteracts the Earth's gravity to a small degree – up to a maximum of 0.3% at the Equator – and reduces the apparent downward acceleration of falling objects.

The second major reason for the difference in gravity at different latitudes is that the Earth's equatorial bulge (itself also caused by centrifugal force from rotation) causes objects at the Equator to be farther from the planet's centre than objects at the poles. Because the force due to gravitational attraction between two bodies (the Earth and the object being weighed) varies inversely with the square of the distance between them, an object at the Equator experiences a weaker gravitational pull than an object at the poles.

In combination, the equatorial bulge and the effects of the surface centrifugal force due to rotation mean that sea-level effective gravity increases from about 9.780 m/s² at the Equator to about 9.832 m/s² at the poles, so an object will weigh about 0.5% more at the poles than at the Equator.

The same two factors influence the direction of the effective gravity (as determined by a plumb line or as the perpendicular to the surface of water in a container). Anywhere on Earth away from the

Equator or poles, effective gravity points not exactly toward the centre of the Earth, but rather perpendicular to the surface of the geoid, which, due to the flattened shape of the Earth, is somewhat toward the opposite pole. About half of the deflection is due to centrifugal force, and half because the extra mass around the Equator causes a change in the direction of the true gravitational force relative to what it would be on a spherical Earth.

Altitude

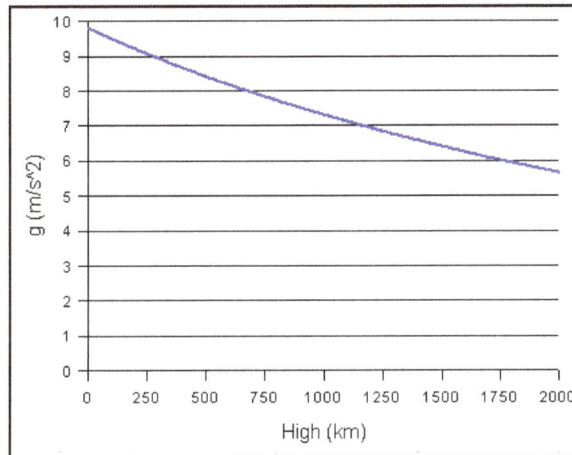

The graph shows the variation in gravity relative to the height of an object.

Gravity decreases with altitude as one rises above the Earth's surface because greater altitude means greater distance from the Earth's centre. All other things being equal, an increase in altitude from sea level to 9,000 metres (30,000 ft) causes a weight decrease of about 0.29%. (An additional factor affecting apparent weight is the decrease in air density at altitude, which lessens an object's buoyancy. This would increase a person's apparent weight at an altitude of 9,000 metres by about 0.08%)

It is a common misconception that astronauts in orbit are weightless because they have flown high enough to escape the Earth's gravity. In fact, at an altitude of 400 kilometres (250 mi), equivalent to a typical orbit of the Space Shuttle, gravity is still nearly 90% as strong as at the Earth's surface. Weightlessness actually occurs because orbiting objects are in free-fall.

The effect of ground elevation depends on the density of the ground. A person flying at 30 000 ft above sea level over mountains will feel more gravity than someone at the same elevation but over the sea. However, a person standing on the earth's surface feels less gravity when the elevation is higher.

The following formula approximates the Earth's gravity variation with altitude:

$$g_h = g_0 \left(\frac{r_e}{r_e + h} \right)^2$$

Where

- g_h is the gravitational acceleration at height h above sea level.

- r_e is the Earth's mean radius.

- g_o is the standard gravitational acceleration.

The formula treats the Earth as a perfect sphere with a radially symmetric distribution of mass; a more accurate mathematical treatment is discussed below.

Depth

An approximate value for gravity at a distance r from the centre of the Earth can be obtained by assuming that the Earth's density is spherically symmetric. The gravity depends only on the mass inside the sphere of radius r. All the contributions from outside cancel out as a consequence of the inverse-square law of gravitation. Another consequence is that the gravity is the same as if all the mass were concentrated at the centre. Thus, the gravitational acceleration at this radius is

$$g(r) = -\frac{GM(r)}{r^2}.$$

where G is the gravitational constant and $M(r)$ is the total mass enclosed within radius r. If the Earth had a constant density ρ, the mass would be $M(r) = (4/3)\pi\rho r^3$ and the dependence of gravity on depth would be

$$g(r) = \frac{4\pi}{3} G\rho r.$$

g at depth d is given by $g'=g(1-d/R)$ where g is acceleration due to gravity on surface of the earth, d is depth and R is radius of Earth. If the density decreased linearly with increasing radius from a density ρ_0 at the centre to ρ_1 at the surface, then $\rho(r) = \rho_0 - (\rho_0 - \rho_1) r / r_e$, and the dependence would be

$$g(r) = \frac{4\pi}{3} G\rho_0 r - \pi G (\rho_0 - \rho_1) \frac{r^2}{r_e}.$$

The actual depth dependence of density and gravity, inferred from seismic travel times, is shown in the graphs below.

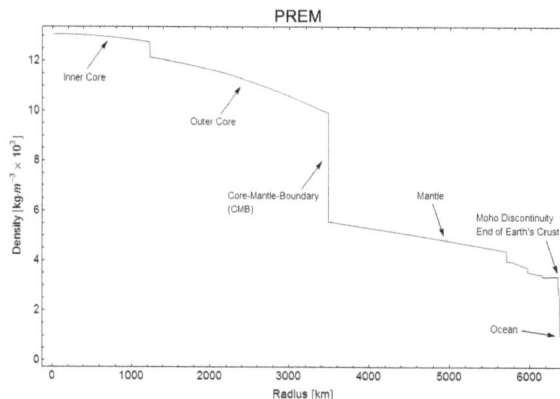

Earth's radial density distribution according to the Preliminary Reference Earth Model (PREM).

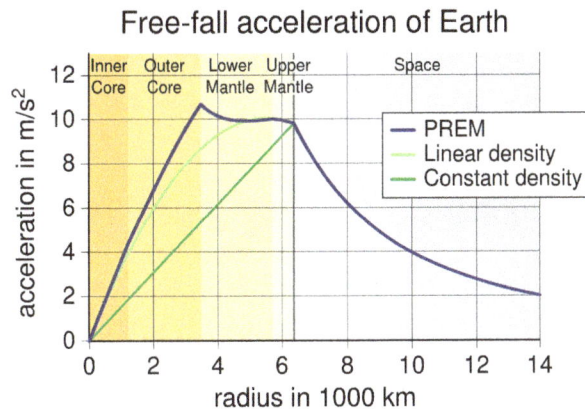

Free-fall acceleration of Earth

Earth's gravity according to the Preliminary Reference Earth Model (PREM). Two models for a spherically symmetric Earth are included for comparison. The dark green straight line is for a constant density equal to the Earth's average density. The light green curved line is for a density that decreases linearly from centre to surface. The density at the centre is the same as in the PREM, but the surface density is chosen so that the mass of the sphere equals the mass of the real Earth.

Local Topography and Geology

Local differences in topography (such as the presence of mountains), geology (such as the density of rocks in the vicinity), and deeper tectonic structure cause local and regional differences in the Earth's gravitational field, known as gravitational anomalies. Some of these anomalies can be very extensive, resulting in bulges in sea level, and throwing pendulum clocks out of synchronisation.

The study of these anomalies forms the basis of gravitational geophysics. The fluctuations are measured with highly sensitive gravimeters, the effect of topography and other known factors is subtracted, and from the resulting data conclusions are drawn. Such techniques are now used by prospectors to find oil and mineral deposits. Denser rocks (often containing mineral ores) cause higher than normal local gravitational fields on the Earth's surface. Less dense sedimentary rocks cause the opposite.

Other Factors

In air, objects experience a supporting buoyancy force which reduces the apparent strength of gravity (as measured by an object's weight). The magnitude of the effect depends on air density (and hence air pressure).

The gravitational effects of the Moon and the Sun (also the cause of the tides) have a very small effect on the apparent strength of Earth's gravity, depending on their relative positions; typical variations are 2 μm/s² (0.2 mGal) over the course of a day.

Comparative Gravities in Various Cities Around the World

The table below shows the gravitational acceleration in various cities around the world. The effect of latitude can be clearly seen with gravity in high-latitude cities: Anchorage (9.826 m/s²), Helsinki (9.825 m/s²), being about 0.5% greater than that in cities near the equator: Kuala Lumpur (9.776 m/s²), Manila(9.780 m/s²). The effect of altitude can be seen in Mexico City (9.776 m/s²; altitude 2,240 metres (7,350 ft)), and by comparing Denver (9.798 m/s²; 1,616 metres (5,302 ft)) with Washington, D.C. (9.801 m/s²; 30 metres (98 ft)), both of which are near 39° N.

Mathematical Models

Latitude Model

If the terrain is at sea level, we can estimate $g\{\phi\}$, the acceleration at latitude ϕ:

$$
\begin{aligned}
g\{\phi\} &= 9.780327\,\mathrm{m\cdot s^{-2}}\left(1+0.0053024\sin^2\phi-0.0000058\sin^2 2\phi\right), \\
&= 9.780327\,\mathrm{m\cdot s^{-2}}\left(1+0.0052792\sin^2\phi+0.0000232\sin^4\phi\right), \\
&= 9.780327\,\mathrm{m\cdot s^{-2}}\left(1.0053024-0.0053256\cos^2\phi+0.0000232\cos^4\phi\right), \\
&= 9.780327\,\mathrm{m\cdot s^{-2}}\left(1.0026454-0.0026512\cos 2\phi+0.0000058\cos^2 2\phi\right)
\end{aligned}
$$

This is the International Gravity Formula 1967, the 1967 Geodetic Reference System Formula, Helmert's equation or Clairaut's formula.

An alternate formula for g as a function of latitude is the WGS (World Geodetic System) 84 Ellipsoidal Gravity Formula:

$$
g\{\phi\} = \mathbb{G}_e\left[\frac{1+k\sin^2\phi}{\sqrt{1-e^2\sin^2\phi}}\right],
$$

where,

- a,b are the equatorial and polar semi-axes, respectively;

- $e^2 = 1-(b/a)^2$ is the spheroid's eccentricity, squared;

- $\mathbb{G}_e, \mathbb{G}_p$ is the defined gravity at the equator and poles, respectively;

- $k = \dfrac{b\mathbb{G}_p - a\mathbb{G}_e}{a\mathbb{G}_e}$ (formula constant);

then, where $\mathbb{G}_p = 9.8321849378\,\mathrm{m\cdot s^{-2}}$,

$$
g\{\phi\} = 9.7803253359\,\mathrm{m\cdot s^{-2}}\left[\frac{1+0.00193185265241\sin^2\phi}{\sqrt{1-0.00669437999013\sin^2\phi}}\right].
$$

The difference between the WGS-84 formula and Helmert's equation is less than 0.68 µm·s⁻².

Free Air Correction

The first correction to be applied to the model is the free air correction (FAC) that accounts for heights above sea level. Near the surface of the Earth (sea level), gravity decreases with height such that linear extrapolation would give zero gravity at a height of one half of the earth's radius -

(9.8 m·s⁻² per 3,200 km.)

Using the mass and radius of the Earth:

$$r_{Earth} = 6.371 \cdot 10^6 \, m$$

$$m_{Earth} = 5.9722 \cdot 10^{24} \, kg$$

The FAC correction factor (Δg) can be derived from the definition of the acceleration due to gravity in terms of G, the Gravitational Constant:

$$g_0 = Gm_{Earth} / r_{Earth}^2 = 9.8196 \frac{m}{s^2}$$

where:

$$G = 6.67384 \cdot 10^{-11} \frac{m^3}{kg \cdot s^2}.$$

At a height h above the nominal surface of the earth g_h is given by:

$$g_h = Gm_{Earth} / (r_{Earth} + h)^2$$

So the FAC for a height h above the nominal earth radius can be expressed:

$$\Delta g_h = \left[Gm_{Earth} / (r_{Earth} + h)^2 \right] - \left[Gm_{Earth} / r_{Earth}^2 \right]$$

This expression can be readily used for programming or inclusion in a spreadsheet. Collecting terms, simplifying and neglecting small terms ($h << r_{Earth}$), however yields the good approximation:

$$\Delta g_h \approx -\frac{Gm_{Earth}}{r_{Earth}^2} \cdot \frac{2h}{r_{Earth}}$$

Using the numerical values above and for a height h in metres:

$$\Delta g_h \approx -3.086 \cdot 10^{-6} h$$

Grouping the latitude and FAC altitude factors the expression most commonly found in the literature is:

$$g\{\phi, h\} = g\{\phi\} - 3.086 \cdot 10^{-6} h$$

where $g\{\phi, h\}$ = acceleration in m·s⁻² at latitude ϕ and altitude h in metres. Alternatively (with the same units for h) the expression can be grouped as follows:

$$g\{\phi,h\} = g\{\phi\} - 3.155 \cdot 10^{-7} h \frac{m}{s^2}$$

Slab Correction

For flat terrain above sea level a second term is added for the gravity due to the extra mass; for this purpose the extra mass can be approximated by an infinite horizontal slab, and we get $2\pi G$ times the mass per unit area, i.e. 4.2×10^{-10} m³·s⁻²·kg⁻¹ (0.042 µGal·kg⁻¹·m²) (the Bouguer correction). For a mean rock density of 2.67 g·cm⁻³ this gives 1.1×10^{-6} s⁻² (0.11 mGal·m⁻¹). Combined with the free-air correction this means a reduction of gravity at the surface of ca. 2 µm·s⁻² (0.20 mGal) for every metre of elevation of the terrain. (The two effects would cancel at a surface rock density of 4/3 times the average density of the whole earth. The density of the whole earth is 5.515 g·cm⁻³, so standing on a slab of something like iron whose density is over 7.35 g·cm⁻³ would increase one's weight.)

For the gravity below the surface we have to apply the free-air correction as well as a double Bouguer correction. With the infinite slab model this is because moving the point of observation below the slab changes the gravity due to it to its opposite. Alternatively, we can consider a spherically symmetrical Earth and subtract from the mass of the Earth that of the shell outside the point of observation, because that does not cause gravity inside. This gives the same result.

Estimating g From the Law of Universal Gravitation

From the law of universal gravitation, the force on a body acted upon by Earth's gravity is given by

$$F = G\frac{m_1 m_2}{r^2} = \left(G\frac{m_1}{r^2}\right)m_2$$

where r is the distance between the centre of the Earth and the body. and here we take m_1 to be the mass of the Earth and m_2 to be the mass of the body.

Additionally, Newton's second law, $F = ma$, where m is mass and a is acceleration, here tells us that

$$F = m_2 g$$

Comparing the two formulas it is seen that:

$$g = G\frac{m_1}{r^2}$$

So, to find the acceleration due to gravity at sea level, substitute the values of the gravitational constant, G, the Earth's mass (in kilograms), m_1, and the Earth's radius (in metres), r, to obtain the value of g:

$$g = G\frac{m_1}{r^2} = 6.67384 \cdot 10^{-11} \text{m}^3 \cdot \text{kg}^{-1} \cdot \text{s}^{-2} \frac{5.9722 \cdot 10^{24} \text{kg}}{(6.371 \cdot 10^6 \text{m})^2} = 9.8196 \text{m} \cdot \text{s}^{-2}$$

Note that this formula only works because of the mathematical fact that the gravity of a uniform spherical body, as measured on or above its surface, is the same as if all its mass were concentrated at a point at its centre. This is what allows us to use the Earth's radius for r.

The value obtained agrees approximately with the measured value of g. The difference may be attributed to several factors, mentioned above under "Variations":

- The Earth is not homogeneous

- The Earth is not a perfect sphere, and an average value must be used for its radius

- This calculated value of g only includes true gravity. It does not include the reduction of constraint force that we perceive as a reduction of gravity due to the rotation of Earth, and some of gravity being counteracted by centrifugal force.

There are significant uncertainties in the values of r and m_1 as used in this calculation, and the value of G is also rather difficult to measure precisely.

If G, g and r are known then a reverse calculation will give an estimate of the mass of the Earth. This method was used by Henry Cavendish.

Free-air Gravity Anomaly

In geophysics, the free-air gravity anomaly, often simply called the free-air anomaly, is the measured gravity anomaly after a *free-air correction* is applied to correct for the elevation at which a measurement is made. The free-air correction does so by adjusting these measurements of gravity to what would have been measured at a reference level. For Earth, this reference level is commonly taken as the mean sea level.

Anomaly

The free-air gravity anomaly is given by the equation:

$$g_F = g_{obs} - g_\lambda + \delta g_F$$

Here, g_F is the free-air gravity anomaly, g_{obs} is observed gravity, g_λ is the correction for latitude (because planetary bodies are not perfect spheres), and   is the free-air correction.

Gravitational acceleration decreases as an inverse square law with the distance at which the measurement is made from the mass. The free air correction is calculated from Newton's Law, as a rate of change of gravity with distance:

$$g = \frac{GM}{R^2}$$

$$\frac{dg}{dR} = -\frac{2GM}{R^3} = -\frac{2g}{R}$$

At the Earth's equator, $2g / R = 0.3086$ mGal/m.

The free-air correction is the amount that must be added to a measurement at height h to correct it to the reference level:

$$\delta g_F = \frac{2g}{R} \times h$$

Here we have assumed that measurements are made relatively close to the surface so that R doesn't vary significantly. Also, there is an assumption that no mass exists between the observation point and the reference level. The Bouguer anomaly and terrain correction are used to account for this.

Theoretical Gravity

In geodesy and geophysics, theoretical gravity or normal gravity is an approximation of the true effective or apparent gravity on Earth's surface by means of a mathematical model representing (a physically smoothed) Earth. The most common model of a smoothed Earth is an Earth ellipsoid, or, more specifically, an Earth spheroid (i.e., an ellipsoid of revolution).

Various, successively more refined, formulas for computing the theoretical gravity are referred to as the International Gravity Formula, the first of which was proposed in 1930 by the International Association of Geodesy. The general shape of that formula is:

$$g(\phi) = g_e \left(1 + A\sin^2(\phi) - B\sin^2(2\phi)\right),$$

in which $g(\varphi)$ is the gravity as a function of the geographic latitude φ of the position whose gravity is to be determined, g_e denotes the gravity at the equator (as determined by measurement), and the coefficients A and B are parameters that must be selected to produce a good global fit to true gravity.

Using the values of the GRS80 reference system, a commonly used specific instantiation of the formula above is given by:

$$g(\phi) = 9.780327\left(1 + 0.0053024\sin^2(\phi) - 0.0000058\sin^2(2\phi)\right)\text{ms}^{-2}$$

Using the appropriate double-angle formula in combination with the Pythagorean identity, this can be rewritten in the equivalent forms

$$g(\phi) = 9.780327\left(1 + 0.0052792\sin^2(\phi) + 0.0000232\sin^4(\phi)\right)\text{ms}^{-2},$$
$$= 9.780327\left(1.0053024 - .0053256\cos^2(\phi) + .0000232\cos^4(\phi)\right)\text{ms}^{-2},$$
$$= 9.780327\left(1.0026454 - 0.0026512\cos(2\phi) + .0000058\cos^2(2\phi)\right)\text{ms}^{-2}.$$

Up to the 1960s, formulas based on the Hayford ellipsoid (1924) and of the famous German geodesist Helmert (1906) were often used. The difference between the semi-major axis (equatorial radius) of the Hayford ellipsoid and that of the modern WGS84 ellipsoid is 251 m; for Helmert's ellipsoid it is only 63 m.

A more recent theoretical formula for gravity as a function of latitude is the International Gravity Formula 1980 (IGF80), also based on the WGS80 ellipsoid but now using the Somigliana equation:

$$g(\phi) = g_e \left[\frac{1 + k \sin^2(\phi)}{\sqrt{1 - e^2 \sin^2(\phi)}} \right],$$

where,

- a, b are the equatorial and polar semi-axes, respectively;

- $e^2 = \dfrac{a^2 - b^2}{a^2}$ is the spheroid's eccentricity, squared;

- g_e, g_p is the defined gravity at the equator and poles, respectively;

- $k = \dfrac{bg_p - ag_e}{ag_e}$ (formula constant);

providing,

$$g(\phi) = 9.7803267715 \left[\frac{1 + 0.001931851353 \sin^2(\phi)}{\sqrt{1 - 0.0066943800229 \sin^2(\phi)}} \right] ms^{-2}$$

A later refinement, based on the WGS84 ellipsoid, is the WGS (World Geodetic System) 1984 Ellipsoidal Gravity Formula:

$$g(\phi) = 9.7803253359 \left[\frac{1 + 0.00193185265241 \sin^2(\phi)}{\sqrt{1 - 0.00669437999013 \sin^2(\phi)}} \right] ms^{-2}.$$

(where $g_p = 9.8321849378 \ ms^{-2} = 9.8321849378 \ ms^{-2}$)

The difference with IGF80 is insignificant when used for geophysical purposes, but may be significant for other uses.

Gravity Gradiometry

Gravity gradiometry is the study and measurement of variations in the acceleration due to gravity. The gravity gradient is the spatial rate of change of gravitational acceleration.

Gravity gradiometry is used by oil and mineral prospectors to measure the density of the subsurface, effectively the rate of change of rock properties. From this information it is possible to build a picture of subsurface anomalies which can then be used to more accurately target oil, gas and mineral deposits. It is also used to image water column density, when locating submerged objects, or determining water depth (bathymetry). Physical scientists use gravimeters to determine the exact size and shape of the earth and they contribute to the gravity compensations applied to inertial navigation systems.

Measuring the Gravity Gradient

Gravity measurements are a reflection of the earth's gravitational attraction, its centripetal force, tidal accelerations due to the sun, moon, and planets, and other applied forces. Gravity gradiometers measure the spatial derivatives of the gravity vector. The most frequently used and intuitive component is the vertical gravity gradient, G_{zz}, which represents the rate of change of vertical gravity (g_z) with height (z). It can be deduced by differencing the value of gravity at two points separated by a small vertical distance, l, and dividing by this distance.

$$G_{zz} = \frac{\partial g_z}{\partial z} \approx \frac{g_z\left(z+\frac{1}{2}\right) - g_z\left(z-\frac{1}{2}\right)}{1}$$

The two gravity measurements are provided by accelerometers which are matched and aligned to a high level of accuracy.

Units

The unit of gravity gradient is the eotvos (abbreviated as E), which is equivalent to 10^{-9} s^{-2} (or 10^{-4} mGal/m). A person walking past at a distance of 2 metres would provide a gravity gradient signal approximately one E. Mountains can give signals of several hundred Eotvos.

Gravity Gradient Tensor

Full tensor gradiometers measure the rate of change of the gravity vector in all three perpendicular directions giving rise to a gravity gradient tensor.

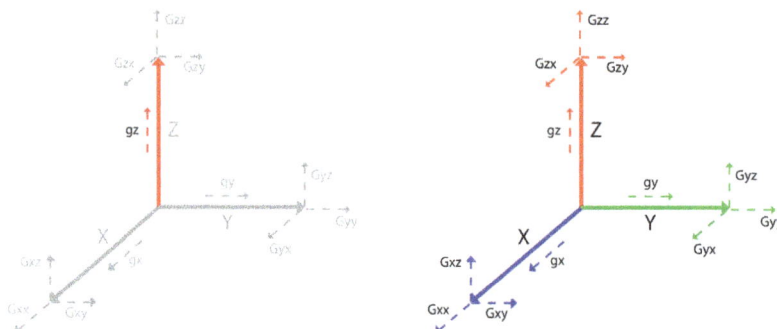

Fig a. Conventional gravity measures ONE component of the gravity field in the vertical direction Gz (LHS), Full tensor gravity gradiometry measures ALL components of the gravity field (RHS).

Comparison to Gravity

Being the derivatives of gravity, the spectral power of gravity gradient signals is pushed to higher frequencies. This generally makes the gravity gradient anomaly more localised to the source than the gravity anomaly. The table (below) and graph (Fig b) compare the g_z and G_{zz} responses from a point source.

	Gravity (g_z)	Gravity gradient (G_{zz})
Signal	$\dfrac{GMz}{\left(r^2+z^2\right)^{\frac{3}{2}}}\times 10^5 \ [\text{mGal}]$	$\dfrac{GM\left(r^2-2z^2\right)}{\left(r^2+z^2\right)^{\frac{5}{2}}}\times 10^9 \ [\text{E}]$
Peak signal ($r = 0$)	$\dfrac{GM}{z^2}\times 10^5$	$\dfrac{2GM}{z^3}\times 10^9$
Full width at half maximum	$1.53z$	$\approx z$
Wavelength (λ)	$3.07z$	$2z$

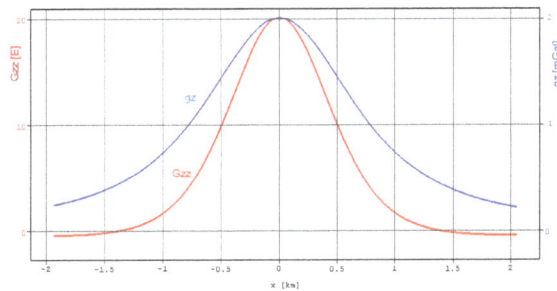

Fig b. Vertical gravity and gravity gradient signals from a point source buried at 1 km depth.

Conversely, gravity measurements have more signal power at low frequency therefore making them more sensitive to regional signals and deeper sources.

Dynamic Survey Environments (Airborne and Marine)

The derivative measurement sacrifices the overall energy in the signal, but significantly reduces the noise due to motional disturbance. On a moving platform, the acceleration disturbance measured by the two accelerometers is the same so that when forming the difference, it cancels in the gravity gradient measurement. This is the principal reason for deploying gradiometers in airborne and marine surveys where the acceleration levels are orders of magnitude greater than the signals of interest. The signal to noise ratio benefits most at high frequency (above 0.01 Hz), where the airborne acceleration noise is largest.

Applications

Gravity gradiometry has predominately been used to image subsurface geology to aid hydrocarbon and mineral exploration. Over 2.5 million line km has now been surveyed using the technique. The surveys highlight gravity anomalies that can be related to geological features such as Salt diapirs, Fault systems, Reef structures, Kimberlite pipes, etc. Other applications include tunnel and bunker detection and the recent GOCE mission that aims to improve the knowledge of ocean circulation.

Gravity Gradiometers

Lockheed Martin Gravity Gradiometers

During the 1970s, as an executive in the Dept. of Defense, John Brett initiated the development of the gravity gradiometer to support the Trident 2 system. A committee was commissioned to seek commercial applications for the Full Tensor Gradient (FTG) system that was developed by Bell Aerospace (later acquired by Lockheed Martin) and was being deployed on US Navy *Ohio*-class Trident submarines designed to aid covert navigation. As the Cold War came to a close, the US Navy released the classified technology and opened the door for full commercialization of the technology. The existence of the gravity gradiometer was famously exposed in the film *The Hunt for Red October* released in 1990.

There are two types of Lockheed Martin gravity gradiometers currently in operation: the 3D Full Tensor Gravity Gradiometer (FTG; deployed in either a fixed wing aircraft or a ship) and the FALCON gradiometer (a partial tensor system with 8 accelerometers and deployed in a fixed wing aircraft or a helicopter). The 3D FTG system contains three gravity gradiometry instruments (GGIs), each consisting of two opposing pairs of accelerometers arranged on a spinning disc with measurement direction in the spin direction.

Other Gravity Gradiometers

Electrostatic gravity gradiometer

> This is the gravity gradiometer deployed on the European Space Agency's GOCE mission. It is a three-axis diagonal gradiometer based on three pairs of electrostatic servo-controlled accelerometers.

ARKeX Exploration gravity gradiometer

> An evolution of technology originally developed for European Space Agency, the Exploration Gravity Gradiometer (EGG), developed by ARKeX (a corporation that is now defunct), uses two key principles of superconductivity to deliver its performance: the Meissner effect, which provides levitation of the EGG proof masses and flux quantization, which gives the EGG its inherent stability. The EGG has been specifically designed for high dynamic survey environments.

Ribbon sensor gradiometer

> The Gravitec gravity gradiometer sensor consists of a single sensing element (a ribbon) that responds to gravity gradient forces. It is designed for borehole applications.

UWA gravity gradiometer

> The University of Western Australia (aka VK-1) Gravity Gradiometer is a superconducting instrument which uses an orthogonal quadrupole responder (OQR) design based on pairs of micro-flexure supported balance beams.

Gedex gravity gradiometer

> The Gedex gravity gradiometer (AKA High-Definition Airborne Gravity Gradiometer, HD-AGG) is also a superconducting OQR-type gravity gradiometer, based on technology developed at the University of Maryland.

Gravimetry

Gravimetry is the measurement of the strength of a gravitational field. Gravimetry may be used when either the magnitude of gravitational field or the properties of matter responsible for its creation are of interest.

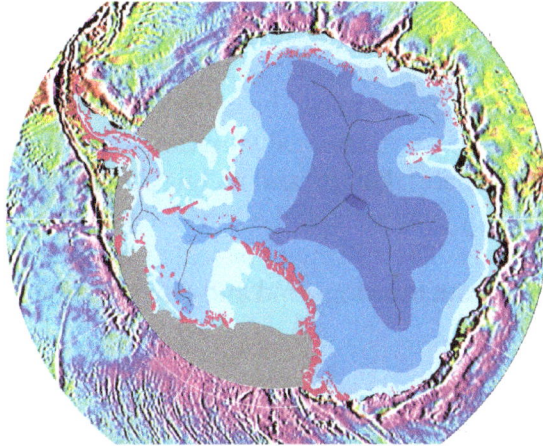

Gravity anomalies covering the Southern Ocean are shown here in false-color relief.
Amplitudes range between -30 mGal (magenta) to +30 mGal (red). This image has been normalized to remove variation due to differences in latitude

Units of Measurement

Gravity is usually measured in units of acceleration. In the SI system of units, the standard unit of acceleration is 1 metre per second squared (abbreviated as m/s²). Other units include the gal (sometimes known as a *galileo*, in either case with symbol Gal), which equals 1 centimetre per second squared, and the *g* (g_n), equal to 9.80665 m/s². The value of the g_n approximately equals the acceleration due to gravity at the Earth's surface (although the value of *g* varies by location).

Measuring Gravity

An instrument used to measure gravity is known as a gravimeter, or gravitometer. For a small body, general relativity predicts gravitational effects indistinguishable from the effects of acceleration by the equivalence principle. Thus, gravimeters can be regarded as special-purpose accelerometers. Many weighing scales may be regarded as simple gravimeters. In one common form, a spring is used to counteract the force of gravity pulling on an object. The change in length of the spring may be calibrated to the force required to balance the gravitational pull. The resulting measurement may be made in units of force (such as the newton), but is more commonly made in units of gals.

Researchers use more sophisticated gravimeters when precise measurements are needed. When measuring the Earth's gravitational field, measurements are made to the precision of microgals to find density variations in the rocks making up the Earth. Several types of gravimeters exist for making these measurements, including some that are essentially refined versions of the spring scale described above. These measurements are used to define gravity anomalies.

Besides precision, stability is also an important property of a gravimeter, as it allows the monitoring of gravity *changes*. These changes can be the result of mass displacements inside the Earth, or of vertical movements of the Earth's crust on which measurements are being made: remember that gravity decreases 0.3 mGal for every metre of height. The study of gravity changes belongs to geodynamics.

The majority of modern gravimeters use specially-designed metal or quartz zero-length springs to support the test mass. Zero-length springs do not follow Hooke's Law, instead they have a force proportional to their length. The special property of these springs is that the natural resonant period of oscillation of the spring-mass system can be made very long - approaching a thousand seconds. This detunes the test mass from most local vibration and mechanical noise, increasing the sensitivity and utility of the gravimeter. Quartz and metal springs are chosen for different reasons; quartz springs are less affected by magnetic and electric fields while metal springs have a much lower drift (elongation) with time. The test mass is sealed in an air-tight container so that tiny changes of barometric pressure from blowing wind and other weather do not change the buoyancy of the test mass in air.

Spring gravimeters are, in practice, relative instruments which measure the difference in gravity between different locations. A relative instrument also requires calibration by comparing instrument readings taken at locations with known complete or absolute values of gravity. Absolute gravimeters provide such measurements by determining the gravitational acceleration of a test mass in vacuum. A test mass is allowed to fall freely inside a vacuum chamber and its position is measured with a laser interferometer and timed with an atomic clock. The laser wavelength is known to ±0.025 ppb and the clock is stable to ±0.03 ppb as well. Great care must be taken to minimize the effects of perturbing forces such as residual air resistance (even in vacuum), vibration, and magnetic forces. Such instruments are capable of an accuracy of about two parts per billion or 0.002 mGal and reference their measurement to atomic standards of length and time. Their primary use is for calibrating relative instruments, monitoring crustal deformation, and in geophysical studies requiring high accuracy and stability. However, absolute instruments are somewhat larger and significantly more expensive than relative spring gravimeters, and are thus relatively rare.

Gravimeters have been designed to mount in vehicles, including aircraft (note the field of aerogravity), ships and submarines. These special gravimeters isolate acceleration from the movement of the vehicle and subtract it from measurements. The acceleration of the vehicles is often hundreds or thousands of times stronger than the changes being measured.

A gravimeter (the *Lunar Surface Gravimeter*) deployed on the surface of the moon during the Apollo 17 mission did not work due to a design error. A second device (the *Traverse Gravimeter Experiment*) functioned as anticipated.

Microgravimetry

Microgravimetry is a rising and important branch developed on the foundation of classical gravimetry. Microgravity investigations are carried out in order to solve various problems of engineering geology, mainly location of voids and their monitoring. Very detailed measurements of high accuracy can indicate voids of any origin, provided the size and depth are large enough to produce gravity effect stronger than is the level of confidence of relevant gravity signal.

History

The modern gravimeter was developed by Lucien LaCoste and Arnold Romberg in 1936.

They also invented most subsequent refinements, including the ship-mounted gravimeter, in 1965, temperature-resistant instruments for deep boreholes, and lightweight hand-carried instruments. Most of their designs remain in use (2005) with refinements in data collection and data processing.

Gravimeter

A gravimeter is an instrument used in gravimetry for measuring the local gravitational field of the Earth. A gravimeter is a type of accelerometer, specialized for measuring the constant downward acceleration of gravity, which varies by about 0.5% over the surface of the Earth. Though the essential principle of design is the same as in other accelerometers, gravimeters are typically designed to be much more sensitive in order to measure very tiny fractional changes within the Earth's gravity of 1 g, caused by nearby geologic structures or the shape of the Earth and by temporal tidal variations. This sensitivity means that gravimeters are susceptible to extraneous vibrations including noise that tend to cause oscillatory accelerations. In practice this is counteracted by integral vibration isolation and signal processing. The constraints on temporal resolution are usually less for gravimeters, so that resolution can be increased by processing the output with a longer time constant. Gravimeters display their measurements in units of gals (cm/s^2), instead of more common units of acceleration.

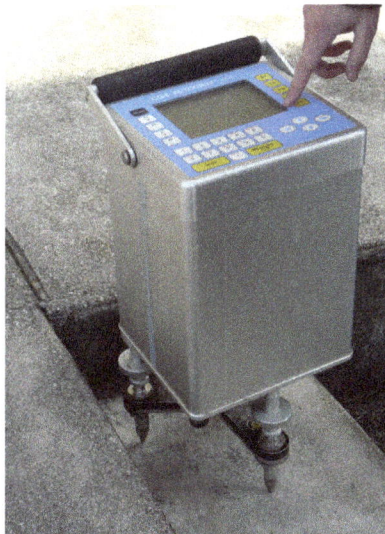

An Autograv CG-5 gravimeter being operated

Gravimeters are used for petroleum and mineral prospecting, seismology, geodesy, geophysical surveys and other geophysical research, and for metrology.

There are two types of gravimeters: relative and absolute. Absolute gravimeters measure the local gravity in absolute units, gals. Relative gravimeters compare the value of gravity at one point with another. They must be calibrated at a location where the gravity is known accurately, and then transported to the location where the gravity is to be measured. They measure the ratio of the gravity at the two points.

Absolute Gravimeters

Absolute gravimeters, which nowadays are made compact so they too can be used in the field, work by directly measuring the acceleration of a mass during free fall in a vacuum, when the accelerometer is rigidly attached to the ground.

The mass includes a retroreflector and terminates one arm of a Michelson interferometer. By counting and timing the interference fringes, the acceleration of the mass can be measured. A more recent development is a "rise and fall" version that tosses the mass upward and measures both upward and downward motion. This allows cancellation of some measurement errors, however "rise and fall" gravimeters are not in common use. Absolute gravimeters are used in the calibration of relative gravimeters, surveying for gravity anomalies (voids), and for establishing the vertical control network.

Relative Gravimeters

Most common *relative* gravimeters are spring-based. They are used in gravity surveys over large areas for establishing the figure of the geoid over those areas. A spring-based relative gravimeter is basically a weight on a spring, and by measuring the amount by which the weight stretches the spring, local gravity can be measured. However, the strength of the spring must be calibrated by placing the instrument in a location with a known gravitational acceleration.

The most accurate relative gravimeters are superconducting gravimeters, which operate by suspending a liquid helium cooled diamagnetic superconducting niobium sphere in an extremely stable magnetic field; the current required to generate the magnetic field that suspends the niobium sphere is proportional to the strength of the Earth's gravitational field. The superconducting gravimeter achieves sensitivities of $10^{-11} ms^{-2}$ (one nanogal), approximately one trillionth (10^{-12}) of the Earth surface gravity. In a demonstration of the sensitivity of the superconducting gravimeter, Virtanen (2006), describes how an instrument at Metsähovi, Finland, detected the gradual increase in surface gravity as workmen cleared snow from its laboratory roof.

Transportable relative gravimeters also exist; they employ an extremely stable inertial platform to compensate for the masking effects of motion and vibration, a difficult engineering feat. The first transportable relative gravimeters were, reportedly, a secret military technology developed in the 1950-60s as a navigational aid for nuclear submarines. Subsequently in the 1980s, transportable relative gravimeters were reverse engineered by the civilian sector for use on ship, then in air and finally satellite borne gravity surveys.

Gravitation of the Moon

The acceleration due to gravity on the surface of the Moon is about 1.625 m/s², about 16.6% that on Earth's surface or 0.16 g. Over the entire surface, the variation in gravitational acceleration is about 0.0253 m/s² (1.6% of the acceleration due to gravity). Because weight is directly dependent upon gravitational acceleration, things on the Moon will weigh only 16.6% of what they weigh on the Earth.

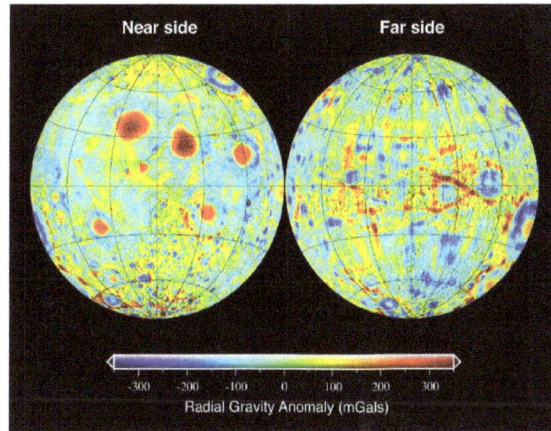

Radial gravity anomaly at the surface of the Moon in Gal (acceleration)

The gravitational field of the Moon has been measured by tracking the radio signals emitted by orbiting spacecraft. The principle used depends on the Doppler effect, whereby the line-of-sight spacecraft acceleration can be measured by small shifts in frequency of the radio signal, and the measurement of the distance from the spacecraft to a station on Earth. Since the gravitational field of the Moon affects the orbit of a spacecraft, one can use these tracking data to detect gravity anomalies. However, because of the Moon's synchronous rotation it is not possible to track spacecraft from Earth much beyond the limbs of the Moon, so until the recent GRAIL mission the farside gravity field was not accurately known.

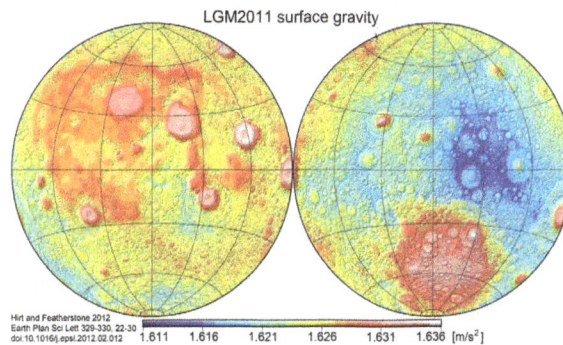

Gravity acceleration at the surface of the Moon in m/s². Near side on the left, far side on the right.
Map from Lunar Gravity Model 2011.

A major feature of the Moon's gravitational field is the presence of mascons, which are large positive gravity anomalies associated with some of the giant impact basins. These anomalies significantly influence the orbit of spacecraft around the Moon, and an accurate gravitational model is necessary in the planning of both manned and unmanned missions. They were initially discovered by the analysis of Lunar Orbiter tracking data: navigation tests prior to the Apollo program showed positioning errors much larger than mission specifications.

Mascons are in part due to the presence of dense mare basaltic lava flows that fill some of the impact basins. However, lava flows by themselves cannot fully explain the gravitational variations, and uplift of the crust-mantle interface is required as well. Based on Lunar Prospector gravitational models, it has been suggested that some mascons exist that do not show evidence for mare basaltic volcanism. The huge expanse of mare basaltic volcanism associated with Oceanus Procellarum does not cause a positive gravity anomaly. The center of gravity of the Moon does not coincide exactly with its geometric center, but is displaced toward the Earth by about 2 kilometers.

Moon – Oceanus Procellarum ("Ocean of Storms")

Ancient rift valleys – rectangular structure (visible – topography – GRAIL gravity gradients) (October 1, 2014).

Ancient rift valleys – context.

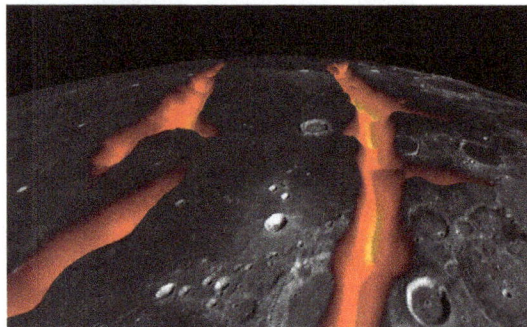

Ancient rift valleys – closeup (artist's concept).

References

- William J. Hinze; Ralph R. B. von Frese; Afif H. Saad (2013). Gravity and Magnetic Exploration: Principles, Practices, and Applications. Cambridge University Press. p. 130. ISBN 978-1-107-32819-8

- A. M. Dziewonski, D. L. Anderson (1981). "Preliminary reference Earth model" (PDF). Physics of the Earth and Planetary Interiors. 25 (4): 297–356. Bibcode:1981PEPI...25..297D. ISSN 0031-9201. doi:10.1016/0031-9201(81)90046-7

- NASA/JPL/University of Texas Center for Space Research. "PIA12146: GRACE Global Gravity Animation". Photojournal. NASA Jet Propulsion Laboratory. Retrieved 30 December 2013

- Tipler, Paul A. (1999). Physics for scientists and engineers. (4th ed.). New York: W.H. Freeman/Worth Publishers. pp. 336–337. ISBN 9781572594913

- Watts, A. B.; Daly, S. F. (May 1981). "Long wavelength gravity and topography anomalies" (PDF). Annual Review of Earth and Planetary Sciences. 9: 415–418. Bibcode:1981AREPS...9..415W. doi:10.1146/annurev. ea.09.050181.002215

- Boynton, Richard (2001). "Precise Measurement of Mass" (PDF). Sawe Paper No. 3147. Arlington, Texas: S.A.W.E., Inc. Retrieved 2007-01-21

- Fowler, C.M.R. (2005). The Solid Earth: An Introduction to Global Geophysics (2 ed.). Cambridge, UK: Cambridge University Press. pp. 205–206. ISBN 0-521-89307-0

- P. Muller; W. Sjogren (1968). "Mascons: Lunar mass concentrations". Science. 161 (3842): 680–684. Bibcode:1968Sci...161..680M. PMID 17801458. doi:10.1126/science.161.3842.680

- Virtanen, H. (2006). Studies of earth dynamics with superconducting gravimeter (PDF). Academic Dissertation at the University of Helsinki, Geodetiska Institutet. Retrieved September 21, 2009

- Lillie, R.J. (1998). Whole Earth Geophysics: An Introductory Textbook for Geologists and Geophysicists. Prentice Hall. ISBN 0-13-490517-2

- Richard A. Kerr (12 April 2013). "The Mystery of Our Moon's Gravitational Bumps Solved?". Science. 340: 128. PMID 23580504. doi:10.1126/science.340.6129.138-a

- C. Hirt; W. E. Featherstone (2012). "A 1.5 km-resolution gravity field model of the Moon". Earth and Planetary Science Letters. 329–330: 22–30. Bibcode:2012E&PSL.329...22H. doi:10.1016/j.epsl.2012.02.012. Retrieved 2012-08-21

- A. Konopliv; S. Asmar; E. Carranza; W. Sjogren; D. Yuan (2001). "Recent gravity models as a result of the Lunar Prospector mission". Icarus. 50: 1–18. Bibcode:2001Icar..150....1K. doi:10.1006/icar.2000.6573

3

Fundamental Concepts of Geodesy

Geodetic datum is a system that is used to locate places on the surface of the Earth. The other concepts related to geodesy are geoid, horizontal position representation, figure of the Earth, metres above sea level and geographical distance. The topics discussed in the chapter are of great importance to broaden the existing knowledge on geodesy.

Geodetic Datum

A geodetic datum or geodetic system is a coordinate system, and a set of reference points, used to locate places on the Earth (or similar objects). An approximate definition of sea level is the datum WGS 84, an ellipsoid, whereas a more accurate definition is Earth Gravitational Model 2008 (EGM2008), using at least 2,159 spherical harmonics. Other datums are defined for other areas or at other times; ED50 was defined in 1950 over Europe and differs from WGS 84 by a few hundred meters depending on where in Europe you look. Mars has no oceans and so no sea level, but at least two martian datums have been used to locate places there.

Datums are used in geodesy, navigation, and surveying by cartographers and satellite navigation systems to translate positions indicated on maps (paper or digital) to their real position on Earth. Each starts with an ellipsoid (stretched sphere), and then defines latitude, longitude and altitude coordinates. One or more locations on the Earth's surface are chosen as anchor "base-points".

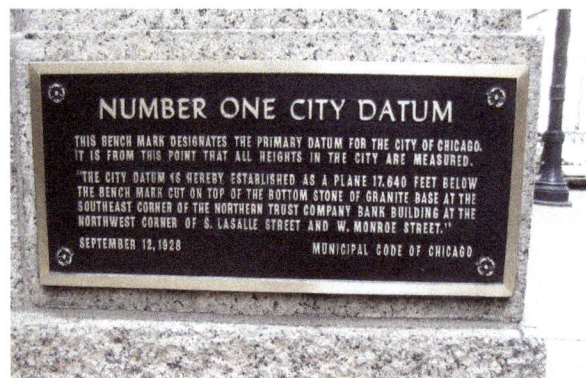

City of Chicago Datum Benchmark

The difference in co-ordinates between datums is commonly referred to as *datum shift*. The datum shift between two particular datums can vary from one place to another within one country or region, and can be anything from zero to hundreds of meters (or several kilometers for some remote islands). The North Pole, South Pole and Equator will be in different positions on different datums, so True North will be slightly different. Different datums use different interpolations for the precise shape and size of the Earth (reference ellipsoids).

Because the Earth is an imperfect ellipsoid, localised datums can give a more accurate representation of the area of coverage than WGS 84. OSGB36, for example, is a better approximation to the geoid covering the British Isles than the global WGS 84 ellipsoid. However, as the benefits of a global system outweigh the greater accuracy, the global WGS 84 datum is becoming increasingly adopted.

Horizontal datums are used for describing a point on the Earth's surface, in latitude and longitude or another coordinate system. Vertical datums measure elevations or depths.

Definition

In surveying and geodesy, a *datum* is a reference system or an approximation of the Earth's surface against which positional measurements are made for computing locations. Horizontal datums are used for describing a point on the Earth's surface, in latitude and longitude or another coordinate system. Vertical datums are used to measure elevations or underwater depths.

Horizontal Datum

The horizontal datum is the model used to measure positions on the Earth. A specific point on the Earth can have substantially different coordinates, depending on the datum used to make the measurement. There are hundreds of local horizontal datums around the world, usually referenced to some convenient local reference point. Contemporary datums, based on increasingly accurate measurements of the shape of the Earth, are intended to cover larger areas. The WGS 84 datum, which is almost identical to the NAD83 datum used in North America and the ETRS89 datum used in Europe, is a common standard datum.

For example, in Sydney there is a 200 metres (700 feet) difference between GPS coordinates configured in GDA (based on global standard WGS 84) and AGD (used for most local maps), which is an unacceptably large error for some applications, such as surveying or site location for scuba diving.

Vertical Datum

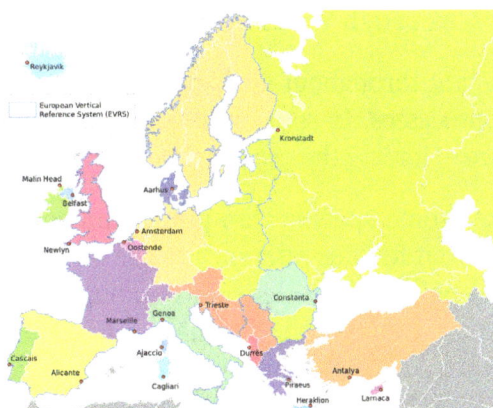

Vertical datums in Europe

A vertical datum is used as a reference point for elevations of surfaces and features on the Earth including terrain, bathymetry, water levels, and man-made structures. Vertical datums are either: tidal, based on sea levels; gravimetric, based on a geoid; or geodetic, based on the same ellipsoid models of the Earth used for computing horizontal datums.

In common usage, elevations are often cited in height above sea level, although what "sea level" actually means is a more complex issue than might at first be thought: the height of the sea surface at any one place and time is a result of numerous effects, including waves, wind and currents, atmospheric pressure, tides, topography, and even differences in the strength of gravity due to the presence of mountains etc.

For the purpose of measuring the height of objects on land, the usual datum used is mean sea level (MSL). This is a tidal datum which is described as the arithmetic mean of the hourly water elevation taken over a specific 19 years cycle. This definition averages out tidal highs and lows (caused by the gravitational effects of the sun and the moon) and short term variations. It will not remove the effects of local gravity strength, and so the height of MSL, relative to a geodetic datum, will vary around the world, and even around one country. Countries tend to choose the mean sea level at one specific point to be used as the standard "sea level" for all mapping and surveying in that country. (For example, in Great Britain, the national vertical datum, Ordnance Datum Newlyn, is based on what was mean sea level at Newlyn in Cornwall between 1915 and 1921). However, zero elevation as defined by one country is not the same as zero elevation defined by another (because MSL is not the same everywhere), which is why locally defined vertical datums differ from one another.

A different principle is used when choosing a datum for nautical charts. For safety reasons, a mariner must be able to know the minimum depth of water that could occur at any point. For this reason, depths and tides on a nautical chart are measured relative to chart datum, which is defined to be a level below which tide rarely falls. Exactly how this is chosen depends on the tidal regime in the area being charted and on the policy of the hydrographic office producing the chart in question; a typical definition is Lowest Astronomical Tide (the lowest tide predictable from the effects of gravity), or Mean Lower Low Water (the average lowest tide of each day), although MSL is sometimes used in waters with very low tidal ranges.

Conversely, if a ship is to safely pass under a low bridge or overhead power cable, the mariner must know the minimum clearance between the masthead and the obstruction, which will occur at high tide. Consequently, bridge clearances etc. are given relative to a datum based on high tide, such as Highest Astronomical Tide or Mean High Water Springs.

Sea level does not remain constant throughout geological time, and so tidal datums are less useful when studying very long-term processes. In some situations sea level does not apply at all — for instance for mapping Mars' surface — forcing the use of a different "zero elevation", such as mean radius.

A geodetic vertical datum takes some specific zero point, and computes elevations based on the geodetic model being used, without further reference to sea levels. Usually, the starting reference point is a tide gauge, so at that point the geodetic and tidal datums might match, but due to sea level variations, the two scales may not match elsewhere. An example of a gravity-based geodetic datum is NAVD88, used in North America, which is referenced to a point in Quebec, Canada. Ellipsoid-based datums such as WGS 84, GRS80 or NAD83 use a theoretical surface that may differ significantly from the geoid.

Geodetic Coordinates

In geodetic coordinates the Earth's surface is approximated by an ellipsoid and locations near the surface are described in terms of latitude (ϕ), longitude (λ) and height (h).

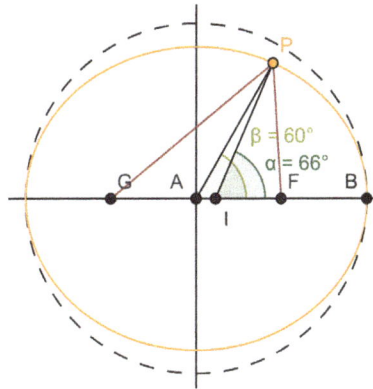

The same position on a spheroid has a different angle for latitude depending on whether the angle is measured from the normal line segment *IP* of the ellipsoid (angle *a*) or the line segment *AP* from the center (angle *β*). Note that the "flatness" of the spheroid (orange) in the image is greater than that of the Earth; as a result, the corresponding difference between the "geodetic" and "geocentric" latitudes is also exaggerated.

Geodetic Versus Geocentric Latitude

It is important to note that geodetic latitude (ϕ) (resp. altitude) is different from geocentric latitude (ϕ') (resp. altitude). Geodetic latitude is determined by the angle between the equatorial plane and normal to the ellipsoid, whereas geocentric latitude is determined by the angle between the equatorial plane and line joining the point to the centre of the ellipsoid. Unless otherwise specified latitude is geodetic latitude.

Earth Reference Ellipsoid

Defining and Derived Parameters

The ellipsoid is completely parameterised by the semi-major axis a and the flattening f.

Parameter	Symbol
Semi-major axis	a
Reciprocal of flattening	$1/f$

From a and f it is possible to derive the semi-minor axis b, first eccentricity e and second eccentricity e' of the ellipsoid

Parameter	Value
semi-minor axis	$b = a(1 - f)$
First eccentricity squared	$e^2 = 1 - b^2/a^2 = 2f - f^2$
Second eccentricity squared	$e'^2 = a^2/b^2 - 1 = f(2 - f)/(1 - f)^2$

Parameters for Some Geodetic Systems

Australian Geodetic Datum 1966 [AGD66] and Australian Geodetic Datum 1984 (AGD84)

AGD66 and AGD84 both use the parameters defined by Australian National Spheroid

Australian National Spheroid (ANS)

ANS Defining Parameters		
Parameter	**Nota-tion**	**Value**
semi-major axis	a	6 378 160.000 m
Reciprocal of Flattening	$1/f$	298.25

Geocentric Datum of Australia 1994 (GDA94)

GDA94 uses the parameters defined by GRS80

Geodetic Reference System 1980 (GRS80)

GRS80 Parameters		
Parameter	**Notation**	**Value**
semi-major axis	a	6 378 137 m
Reciprocal of flattening	$1/f$	298.257 222 101

World Geodetic System 1984 (WGS 84)

The Global Positioning System (GPS) uses the World Geodetic System 1984 (WGS 84) to determine the location of a point near the surface of the Earth.

WGS 84 Defining Parameters		
Parameter	**Notation**	**Value**
semi-major axis	a	6 378 137.0 m
Reciprocal of flattening	$1/f$	298.257 223 563

WGS 84 derived geometric constants		
Constant	**Notation**	**Value**
Semi-minor axis	b	6 356 752.3142 m
First eccentricity squared	e^2	$6.694\ 379\ 990\ 14 \times 10^{-3}$
Second eccentricity squared	e'^2	$6.739\ 496\ 742\ 28 \times 10^{-3}$

A more comprehensive list of geodetic systems can be found here

Conversion Calculations

Datum conversion is the process of converting the coordinates of a point from one datum system to another. Datum conversion may frequently be accompanied by a change of grid projection.

Reference Datums

A reference datum is a known and constant surface which is used to describe the location of unknown points on the Earth. Since reference datums can have different radii and different center points, a specific point on the Earth can have substantially different coordinates depending on the datum used to make the measurement. There are hundreds of locally developed reference datums around the world, usually referenced to some convenient local reference point. Contemporary datums, based on increasingly accurate measurements of the shape of the Earth, are intended to cover larger areas. The most common reference Datums in use in North America are NAD27, NAD83, and WGS 84.

The North American Datum of 1927 (NAD 27) is "the horizontal control datum for the United States that was defined by a location and azimuth on the Clarke spheroid of 1866, with origin at (the survey station) Meades Ranch (Kansas)." ... The geoidal height at Meades Ranch was assumed to be zero, as sufficient gravity data was not available, and this was needed to relate surface measurements to the datum. "Geodetic positions on the North American Datum of 1927 were derived from the (coordinates of and an azimuth at Meades Ranch) through a readjustment of the triangulation of the entire network in which Laplace azimuths were introduced, and the Bowie method was used." NAD27 is a local referencing system covering North America.

The North American Datum of 1983 (NAD 83) is "The horizontal control datum for the United States, Canada, Mexico, and Central America, based on a geocentric origin and the Geodetic Reference System 1980 (GRS80). "This datum, designated as NAD 83 ...is based on the adjustment of 250,000 points including 600 satellite Doppler stations which constrain the system to a geocentric origin." NAD83 may be considered a local referencing system.

WGS 84 is the World Geodetic System of 1984. It is the reference frame used by the U.S. Department of Defense (DoD) and is defined by the National Geospatial-Intelligence Agency (NGA) (formerly the Defense Mapping Agency, then the National Imagery and Mapping Agency). WGS 84 is used by DoD for all its mapping, charting, surveying, and navigation needs, including its GPS "broadcast" and "precise" orbits. WGS 84 was defined in January 1987 using Doppler satellite surveying techniques. It was used as the reference frame for broadcast GPS Ephemerides (orbits) beginning January 23, 1987. At 0000 GMT January 2, 1994, WGS 84 was upgraded in accuracy using GPS measurements. The formal name then became WGS 84 (G730), since the upgrade date coincided with the start of GPS Week 730. It became the reference frame for broadcast orbits on June 28, 1994. At 0000 GMT September 30, 1996 (the start of GPS Week 873), WGS 84 was redefined again and was more closely aligned with International Earth Rotation Service (IERS) frame ITRF 94. It was then formally called WGS 84 (G873). WGS 84 (G873) was adopted as the reference frame for broadcast orbits on January 29, 1997. Another update brought it to WGS84(G1674).

The WGS 84 datum, within two meters of the NAD83 datum used in North America, is the only world referencing system in place today. WGS 84 is the default standard datum for coordinates stored in recreational and commercial GPS units.

Users of GPS are cautioned that they must always check the datum of the maps they are using. To correctly enter, display, and to store map related map coordinates, the datum of the map must be entered into the GPS map datum field.

Geoid

The geoid is the shape that the surface of the oceans would take under the influence of Earth's gravity and rotation alone, in the absence of other influences such as winds and tides. This surface is extended through the continents (such as with very narrow hypothetical canals). All points on a geoid surface have the same gravitational potential energy (the sum of gravitational potential energy and centrifugal potential energy). The geoid can be defined at any value of gravitational potential such as within the Earth's crust or far out in space, not just at sea level. The force of gravity acts everywhere perpendicular to the geoid, meaning that plumb lines point perpendicular and water levels parallel to the geoid if only gravity and rotational acceleration were at work.

Specifically, the geoid is the equipotential surface that would coincide with the mean ocean surface of Earth if the oceans and atmosphere were in equilibrium, at rest relative to the rotating Earth, and extended through the continents (such as with very narrow canals). According to Gauss, who first described it, it is the "mathematical figure of Earth", a smooth but highly irregular surface whose shape results from the uneven distribution of mass within and on the surface of Earth. It does not correspond to the actual surface of Earth's crust, but to a surface which can only be known through extensive gravitational measurements and calculations. Despite being an important concept for almost two hundred years in the history of geodesy and geophysics, it has only been defined to high precision since advances in satellite geodesy in the late 20th century. It is often described as the true physical figure of the Earth, in contrast to the idealized geometrical figure of a reference ellipsoid.

The surface of the geoid is higher than the reference ellipsoid wherever there is a positive gravity anomaly (mass excess) and lower than the reference ellipsoid wherever there is a negative gravity anomaly (mass deficit).

Description

Deviation of the Geoid from the idealized figure of the Earth
(difference between the EGM96 geoid and the WGS84 reference ellipsoid)

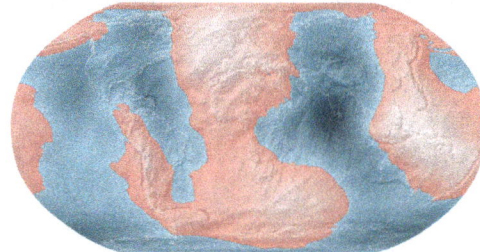

Red areas are above the idealized ellipsoid; blue areas are below.

-107.0 m   0 m   +85.4 m

Map of the undulation of the geoid, in meters (based on the EGM96 gravity model and the WGS84 reference ellipsoid).

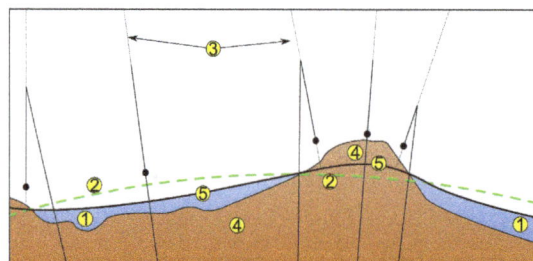

1. Ocean 2. Reference ellipsoid 3. Local plumb line 4. Continent 5. Geoid

The geoid surface is irregular, unlike the reference ellipsoid which is a mathematical idealized representation of the physical Earth, but considerably smoother than Earth's physical surface. Although the physical Earth has excursions of +8,848 m (Mount Everest) and −429 m (Dead Sea), the geoid's variation ranges from −106 to +85 m, less than 200 m total compared to a perfect mathematical ellipsoid.

If the ocean surface were isopycnic (of constant density) and undisturbed by tides, currents, or weather, it would closely approximate the geoid. The permanent deviation between the geoid and mean sea level is called ocean surface topography. If the continental land masses were crisscrossed by a series of tunnels or canals, the sea level in these canals would also very nearly coincide with the geoid. In reality the geoid does not have a physical meaning under the continents, but geodesists are able to derive the heights of continental points above this imaginary, yet physically defined, surface by a technique called spirit leveling.

Being an equipotential surface, the geoid is by definition a surface to which the force of gravity is everywhere perpendicular. This means that when traveling by ship, one does not notice the undulations of the geoid; the local vertical (plumb line) is always perpendicular to the geoid and the local horizon tangential to it. Likewise, spirit levels will always be parallel to the geoid.

On a long voyage, spirit leveling indicates height variations even though the ship is always at sea level (neglecting the effects of tides). This is because GPS satellites, orbiting about the center of gravity of the Earth, can only measure heights relative to a geocentric reference ellipsoid. To obtain one's geoidal height, a raw GPS reading must be corrected. Conversely, height determined by spirit leveling from a tidal measurement station, as in traditional land surveying, will always be geoidal height. Modern GPS receivers have a grid implemented inside where they obtain the geoid (e.g. EGM-96) height over the World Geodetic System (WGS) ellipsoid from the current position. Then they are able to correct the height above WGS ellipsoid to the height above WGS84 geoid. When height is not zero on a ship, the discrepancy is due to other factors such as ocean tides, atmospheric pressure (meteorological effects) and local sea surface topography.

Simplified Example

The gravitational field of the earth is neither perfect nor uniform. A flattened ellipsoid is typically used as the idealized earth, but even if the earth were perfectly spherical, the strength of gravity would not be the same everywhere, because density (and therefore mass) varies throughout the planet. This is due to magma distributions, mountain ranges, deep sea trenches, and so on.

If that perfect sphere were then covered in water, the water would not be the same height everywhere. Instead, the water level would be higher or lower depending on the particular strength of gravity in that location.

Spherical Harmonics Representation

Spherical harmonics are often used to approximate the shape of the geoid. The current best such set of spherical harmonic coefficients is EGM96 (Earth Gravity Model 1996), determined in an international collaborative project led by NIMA. The mathematical description of the non-rotating part of the potential function in this model is:

$$V = \frac{GM}{r}\left(1 + \sum_{n=2}^{n_{max}}\left(\frac{a}{r}\right)^n \sum_{m=0}^{n}\overline{P}_{nm}(\sin\phi)\left[\overline{C}_{nm}\cos m\lambda + \overline{S}_{nm}\sin m\lambda\right]\right),$$

where ϕ and λ are *geocentric* (spherical) latitude and longitude respectively, $\bar{P}_{nm}$ are the fully normalized associated Legendre polynomials of degree n and order m, and $\bar{C}_{nm}$ and $\bar{S}_{nm}$ are the numerical coefficients of the model based on measured data. Note that the above equation describes the Earth's gravitational potential V, not the geoid itself, at location ϕ, λ, r, the co-ordinate r being the *geocentric radius*, i.e., distance from the Earth's centre. The geoid is a particular equipotential surface, and is somewhat involved to compute. The gradient of this potential also provides a model of the gravitational acceleration. EGM96 contains a full set of coefficients to degree and order 360 (i.e. $n_{max} = 360$), describing details in the global geoid as small as 55 km (or 110 km, depending on your definition of resolution). The number of coefficients, $\bar{C}_{nm}$ and $\bar{S}_{nm}$, can be determined by first observing in the equation for V that for a specific value of n there are two coefficients for every value of m except for m = 0. There is only one coefficient when m=0 since $\sin(0\lambda) = 0$. There are thus (2n+1) coefficients for every value of n. Using these facts and the formula, $\sum_{I=1}^{L} I = L(L+1)/2$, it follows that the total number of coefficients is given by

$$\sum_{n=2}^{n_{max}} (2n+1) = n_{max}(n_{max}+1) + n_{max} - 3 = 130317 \text{ using the EGM96 value of } n_{max} = 360.$$

For many applications the complete series is unnecessarily complex and is truncated after a few (perhaps several dozen) terms.

New even higher resolution models are currently under development. For example, many of the authors of EGM96 are working on an updated model that should incorporate much of the new satellite gravity data. and should support up to degree and order 2160 (1/6 of a degree, requiring over 4 million coefficients).

NGA has announced the availability of EGM2008, complete to spherical harmonic degree and order 2159, and contains additional coefficients extending to degree 2190 and order 2159. Software and data is on the Earth Gravitational Model 2008.

Precise Geoid

The Precise Geoid Solution by Vaníček and co-workers improved on the Stokesian approach to geoid computation. Their solution enables millimetre-to-centimetre accuracy in geoid computation, an order-of-magnitude improvement from previous classical solutions.

Causes for Geoid Anomalies

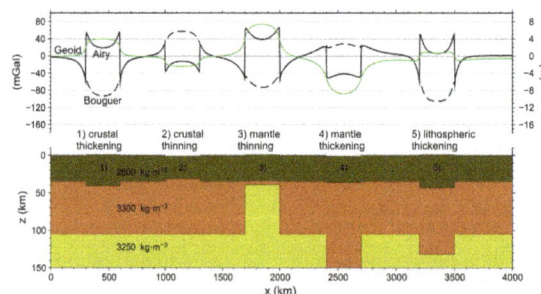

Gravity and Geoid anomalies caused by various crustal and lithospheric thickness changes relative to a reference configuration. All settings are under local isostatic compensation.

Variations in the height of the geoidal surface are related to density anomalous distributions within the Earth. Geoid measures help thus to understand the internal structure of the planet. Synthetic calculations show that the geoidal signature of a thickened crust (for example, in orogenic belts produced by continental collision) is positive, opposite to what should be expected if the thickening affects the entire lithosphere.

Time-variability

Recent satellite missions, such as GOCE and GRACE, have enabled the study of time-variable geoid signals. The first products based on GOCE satellite data became available online in June 2010, through the European Space Agency (ESA)'s Earth observation user services tools. ESA launched the satellite in March 2009 on a mission to map Earth's gravity with unprecedented accuracy and spatial resolution. On 31 March 2011, the new geoid model was unveiled at the Fourth International GOCE User Workshop hosted at the Technische Universität München in Munich, Germany. Studies using the time-variable geoid computed from GRACE data have provided information on global hydrologic cycles, mass balances of ice sheets, and postglacial rebound. From postglacial rebound measurements, time-variable GRACE data can be used to deduce the viscosity of Earth's mantle.

Celestial Bodies

The concept of the geoid has been extended to other planets and also moons, as well as asteroids.

The geoid of Mars has been measured using flight paths of satellite missions such as Mariner 9 and Viking. The main departures from the ellipsoid expected of an ideal fluid are from the Tharsis volcanic plateau, a continent-size region of elevated terrain, and its antipodes.

Geographic Coordinate System

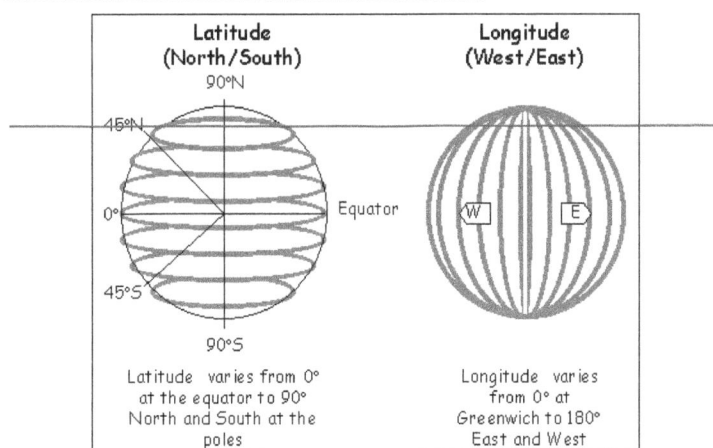

Latitude (North/South)	Longitude (West/East)
90°N	
45°N	
0° — Equator	W E
45°S	
90°S	
Latitude varies from 0° at the equator to 90° North and South at the poles	Longitude varies from 0° at Greenwich to 180° East and West

Longitude lines are perpendicular to and latitude lines are parallel to the equator.

A geographic coordinate system is a coordinate system used in geography that enables every location on Earth to be specified by a set of numbers, letters or symbols. The coordinates are often chosen such that one of the numbers represents a vertical position, and two or three of the numbers represent a horizontal position. A common choice of coordinates is latitude, longitude and elevation.

To specify a location on a two-dimensional map requires a map projection.

History

The invention of a geographic coordinate system is generally credited to Eratosthenes of Cyrene, who composed his now-lost *Geography* at the Library of Alexandria in the 3rd century BC. A century later, Hipparchus of Nicaea improved on this system by determining latitude from stellar measurements rather than solar altitude and determining longitude by using simultaneous timings of lunar eclipses, rather than dead reckoning. In the 1st or 2nd century, Marinus of Tyre compiled an extensive gazetteer and mathematically-plotted world map using coordinates measured east from a prime meridian at the westernmost known land, designated the Fortunate Isles, off the coast of western Africa around the Canary or Cape Verde Islands, and measured north or south of the island of Rhodes off Asia Minor. Ptolemy credited him with the full adoption of longitude and latitude, rather than measuring latitude in terms of the length of the midsummer day. Ptolemy's 2nd-century *Geography* used the same prime meridian but measured latitude from the equator instead. After their work was translated into Arabic in the 9th century, Al-Khwārizmī's *Book of the Description of the Earth* corrected Marinus' and Ptolemy's errors regarding the length of the Mediterranean Sea, causing medieval Arabic cartography to use a prime meridian around 10° east of Ptolemy's line. Mathematical cartography resumed in Europe following Maximus Planudes' recovery of Ptolemy's text a little before 1300; the text was translated into Latin at Florence by Jacobus Angelus around 1407.

In 1884, the United States hosted the International Meridian Conference, attended by representatives from twenty-five nations. Twenty-two of them agreed to adopt the longitude of the Royal Observatory in Greenwich, England as the zero-reference line. The Dominican Republic voted against the motion, while France and Brazil abstained. France adopted Greenwich Mean Time in place of local determinations by the Paris Observatory in 1911.

Geographic Latitude and Longitude

Equator

The "latitude" (abbreviation: Lat., φ, or phi) of a point on Earth's surface is the angle between the equatorial plane and the straight line that passes through that point and through (or close to) the center of the Earth. Lines joining points of the same latitude trace circles on the surface of Earth called parallels, as they are parallel to the equator and to each other. The north pole is 90° N; the south pole is 90° S. The 0° parallel of latitude is designated the equator, the fundamental plane of all geographic coordinate systems. The equator divides the globe into Northern and Southern Hemispheres.

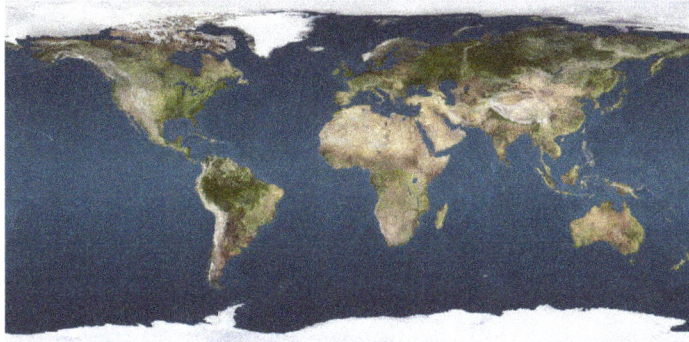

Prime Meridian

The "longitude" (abbreviation: Long., λ, or lambda) of a point on Earth's surface is the angle east or west of a reference meridian to another meridian that passes through that point. All meridians are halves of great ellipses (often called great circles), which converge at the north and south poles. The meridian of the British Royal Observatory in Greenwich, in south-east London, England, is the international prime meridian, although some organizations—such as the French Institut Géographique National—continue to use other meridians for internal purposes. The prime meridian determines the proper Eastern and Western Hemispheres, although maps often divide these hemispheres further west in order to keep the Old World on a single side. The antipodal meridian of Greenwich is both 180°W and 180°E. This is not to be conflated with the International Date Line, which diverges from it in several places for political reasons, including between far eastern Russia and the far western Aleutian Islands.

The combination of these two components specifies the position of any location on the surface of Earth, without consideration of altitude or depth. The grid formed by lines of latitude and longitude is known as a "graticule". The origin/zero point of this system is located in the Gulf of Guinea about 625 km (390 mi) south of Tema, Ghana.

Complexity of the Problem

To completely specify a location of a topographical feature on, in, or above Earth, one also has to specify the vertical distance from Earth's center or surface.

Earth is not a sphere, but an irregular shape approximating a biaxial ellipsoid. It is nearly spherical, but has an equatorial bulge making the radius at the equator about 0.3% larger than the radius measured through the poles. The shorter axis approximately coincides with the axis of rotation. Though early navigators thought of the sea as a flat surface that could be used as a vertical datum, this is not actually the case. Earth has a series of layers of equal potential energy within its gravitational field. Height is a measurement at right angles to this surface, roughly toward Earth's centre, but local variations make the equipotential layers irregular (though roughly ellipsoidal). The choice of which layer to use for defining height is arbitrary.

Common Baselines

Common height baselines include

- The surface of the datum ellipsoid, resulting in an *ellipsoidal height*

- The mean sea level as described by the gravity geoid, yielding the orthometric height

- A vertical datum, yielding a dynamic height relative to a known reference height.

Along with the latitude ϕ and longitude λ, the height h provides the three-dimensional *geodetic coordinates* or *geographic coordinates* for a location.

Datums

In order to be unambiguous about the direction of "vertical" and the "surface" above which they are measuring, map-makers choose a reference ellipsoid with a given origin and orientation that best fits their need for the area they are mapping. They then choose the most appropriate mapping of the spherical coordinate system onto that ellipsoid, called a terrestrial reference system or geodetic datum.

Datums may be global, meaning that they represent the whole earth, or they may be local, meaning that they represent an ellipsoid best-fit to only a portion of the earth. Points on the earth's surface move relative to each other due to continental plate motion, subsidence, and diurnal movement caused by the moon and the tides. This daily movement can be as much as a metre. Continental movement can be up to 10 cm a year, or 10 m in a century. A weather system high-pressure area can cause a sinking of 5 mm. Scandinavia is rising by 1 cm a year as a result of the melting of the ice sheets of the last ice age, but neighbouring Scotland is rising by only 0.2 cm. These changes are insignificant if a local datum is used, but are statistically significant if a global datum is used.

Examples of global datums include World Geodetic System (WGS 84), the default datum used for the Global Positioning System, and the International Terrestrial Reference Frame (ITRF), used for estimating continental drift and crustal deformation. The distance to Earth's centre can be used both for very deep positions and for positions in space.

Local datums chosen by a national cartographical organisation include the North American Datum, the European ED50, and the British OSGB36. Given a location, the datum provides the latitude ϕ and longitude λ. In the United Kingdom there are three common latitude, longitude, and height systems in use. WGS 84 differs at Greenwich from the one used on published maps OSGB36 by approximately 112m. The military system ED50, used by NATO, differs from about 120m to 180m.

The latitude and longitude on a map made against a local datum may not be the same as one obtained from a GPS receiver. Coordinates from the mapping system can sometimes be roughly changed into another datum using a simple translation. For example, to convert from ETRF89 (GPS) to the Irish Grid add 49 metres to the east, and subtract 23.4 metres from the north. More generally one datum is changed into any other datum using a process called Helmert transformations. This involves converting the spherical coordinates into Cartesian coordinates and applying a seven parameter transformation (translation, three-dimensional rotation), and converting back.

In popular GIS software, data projected in latitude/longitude is often represented as a 'Geographic Coordinate System'. For example, data in latitude/longitude if the datum is the North American Datum of 1983 is denoted by 'GCS North American 1983'.

Map Projection

To establish the position of a geographic location on a map, a map projection is used to convert geodetic coordinates to two-dimensional coordinates on a map; it projects the datum ellipsoidal coordinates and height onto a flat surface of a map. The datum, along with a map projection applied to a grid of reference locations, establishes a *grid system* for plotting locations. Common map projections in current use include the Universal Transverse Mercator (UTM), the Military Grid Reference System (MGRS), the United States National Grid (USNG), the Global Area Reference System (GARS) and the World Geographic Reference System (GEO-REF). Coordinates on a map are usually in terms northing N and easting E offsets relative to a specified origin.

Map projection formulas depend in the geometry of the projection as well as parameters dependent on the particular location at which the map is projected. The set of parameters can vary based on type of project and the conventions chosen for the projection. For the transverse Mercator projection used in UTM, the parameters associated are the latitude and longitude of the natural origin, the false northing and false easting, and an overall scale factor. Given the parameters associated with particular location or grin, the projection formulas for the transverse Mercator are a complex mix of algebraic and trigonometric functions.

UTM and UPS Systems

The Universal Transverse Mercator (UTM) and Universal Polar Stereographic (UPS) coordinate systems both use a metric-based cartesian grid laid out on a conformally projected surface to locate positions on the surface of the Earth. The UTM system is not a single map projection but a series of sixty, each covering 6-degree bands of longitude. The UPS system is used for the polar regions, which are not covered by the UTM system.

Stereographic Coordinate System

During medieval times, the stereographic coordinate system was used for navigation purposes. The stereographic coordinate system was superseded by the latitude-longitude system. Although no longer used in navigation, the stereographic coordinate system is still used in modern times to describe crystallographic orientations in the fields of crystallography, mineralogy and materials science.

Cartesian Coordinates

Every point that is expressed in ellipsoidal coordinates can be expressed as an rectilinear x y z

(Cartesian) coordinate. Cartesian coordinates simplify many mathematical calculations. The Cartesian systems of different datums are not equivalent.

Earth-centered, Earth-fixed

The earth-centered earth-fixed (also known as the ECEF, ECF, or conventional terrestrial coordinate system) rotates with the Earth and has its origin at the center of the Earth.

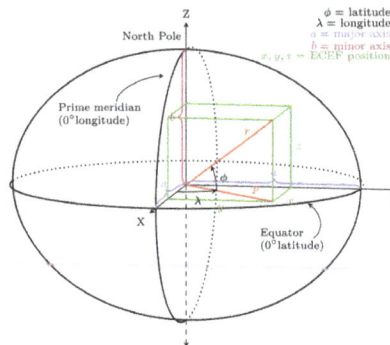

Earth Centered, Earth Fixed coordinates in relation to latitude and longitude.

The conventional right-handed coordinate system puts:

- The origin at the center of mass of the earth, a point close to the Earth's center of figure

- The Z axis on the line between the north and south poles, with positive values increasing northward (but does not exactly coincide with the Earth's rotational axis)

- The X and Y axes in the plane of the equator

- The X axis passing through extending from 180 degrees longitude at the equator (negative) to 0 degrees longitude (prime meridian) at the equator (positive)

- The Y axis passing through extending from 90 degrees west longitude at the equator (negative) to 90 degrees east longitude at the equator (positive)

An example is the NGS data for a brass disk near Donner Summit, in California. Given the dimensions of the ellipsoid, the conversion from lat/lon/height-above-ellipsoid coordinates to X-Y-Z is straightforward—calculate the X-Y-Z for the given lat-lon on the surface of the ellipsoid and add the X-Y-Z vector that is perpendicular to the ellipsoid there and has length equal to the point's height above the ellipsoid. The reverse conversion is harder: given X-Y-Z we can immediately get longitude, but no closed formula for latitude and height exists. Using Bowring's formula in 1976 *Survey Review* the first iteration gives latitude correct within 10^{-11} degree as long as the point is within 10000 meters above or 5000 meters below the ellipsoid.

Local East, North, up (ENU) Coordinates

In many targeting and tracking applications the local East, North, Up (ENU) Cartesian coordinate system is far more intuitive and practical than ECEF or Geodetic coordinates. The local ENU coordinates are formed from a plane tangent to the Earth's surface fixed to a specific location and hence it is sometimes known as a "Local Tangent" or "local geodetic" plane. By convention the east axis is labeled x, the north y and the up z.

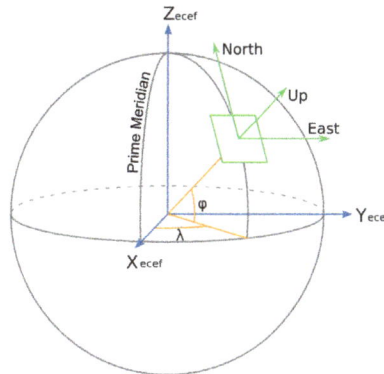

Earth Centred Earth Fixed and East, North, Up coordinates.

Local North, East, Down (NED) Coordinates

Also known as local tangent plane (LTP). In an airplane, most objects of interest are below the aircraft, so it is sensible to define down as a positive number. The North, East, Down (NED) coordinates allow this as an alternative to the ENU local tangent plane. By convention, the north axis is labeled x', the east y' and the down z'. To avoid confusion between x and x', etc. in this web page we will restrict the local coordinate frame to ENU.

Expressing Latitude and Longitude as Linear Units

On the GRS80 or WGS84 spheroid at sea level at the equator, one latitudinal second measures *30.715 metres*, one latitudinal minute is *1843 metres* and one latitudinal degree is *110.6 kilometres*. The circles of longitude, meridians, meet at the geographical poles, with the west-east width of a second naturally decreasing as latitude increases. On the equator at sea level, one longitudinal second measures *30.92 metres*, a longitudinal minute is *1855 metres* and a longitudinal degree is *111.3 kilometres*. At 30° a longitudinal second is *26.76 metres*, at Greenwich (51°28′38″N) *19.22 metres*, and at 60° it is *15.42 metres*.

On the WGS84 spheroid, the length in meters of a degree of latitude at latitude φ (that is, the distance along a north-south line from latitude (φ − 0.5) degrees to (φ + 0.5) degrees) is about

$$111132.92 - 559.82\cos 2\varphi + 1.175\cos 4\varphi - 0.0023\cos 6\varphi$$

Similarly, the length in meters of a degree of longitude can be calculated as

$$111412.84\cos \varphi - 93.5\cos 3\varphi + 0.118\cos 5\varphi$$

(Those coefficients can be improved, but as they stand the distance they give is correct within a centimeter.)

An alternative method to estimate the length of a longitudinal degree at latitude φ is to assume a spherical Earth (to get the width per minute and second, divide by 60 and 3600, respectively):

$$\frac{\pi}{180} M_r \cos\varphi$$

where Earth's average meridional radius M_r is 6,367,449 m. Since the Earth is not spherical that result can be off by several tenths of a percent; a better approximation of a longitudinal degree at latitude φ is

$$\frac{\pi}{180} a \cos\beta$$

where Earth's equatorial radius a equals *6,378,137 m* and $\tan\beta = \frac{b}{a}\tan\varphi$; ; for the GRS80 and WGS84 spheroids, b/a calculates to be 0.99664719. (β is known as the reduced (or parametric) latitude). Aside from rounding, this is the exact distance along a parallel of latitude; getting the distance along the shortest route will be more work, but those two distances are always within 0.6 meter of each other if the two points are one degree of longitude apart.

Longitudinal length equivalents at selected latitudes					
Latitude	**City**	**Degree**	**Minute**	**Second**	**±0.0001°**
60°	Saint Petersburg	55.80 km	0.930 km	15.50 m	5.58 m
51° 28′ 38″ N	Greenwich	69.47 km	1.158 km	19.30 m	6.95 m
45°	Bordeaux	78.85 km	1.31 km	21.90 m	7.89 m
30°	New Orleans	96.49 km	1.61 km	26.80 m	9.65 m
0°	Quito	111.3 km	1.855 km	30.92 m	11.13 m

Geostationary Coordinates

Geostationary satellites (e.g., television satellites) are over the equator at a specific point on Earth, so their position related to Earth is expressed in longitude degrees only. Their latitude is always zero (or approximately so), that is, over the equator.

On other Celestial Bodies

Similar coordinate systems are defined for other celestial bodies such as:

- A similarly well-defined system based on the reference ellipsoid for Mars.
- Selenographic coordinates for the Moon

Horizontal Position Representation

A position representation is the parameters used to express a position relative to a reference. Representing position in three dimensions is often done by a Euclidean vector. However, when representing position relative to the Earth it is often more convenient to represent vertical position as altitude or depth, and to use some other parameters to represent horizontal position. There are also several applications where only the horizontal position is of interest, this might e.g. be the case for ships and ground vehicles/cars.

There are several options for horizontal position representations, each with different properties which makes them appropriate for different applications. Latitude/longitude and UTM are common horizontal position representations.

The horizontal position has two degrees of freedom, and thus two parameters are sufficient to uniquely describe such a position. However, similarly to the use of Euler angles as a formalism for representing rotations, using only the minimum number of parameters gives singularities, and thus three parameters are required for the horizontal position to avoid this.

Latitude and Longitude

The most common horizontal position representation is latitude and longitude. The parameters are intuitive and well known, and are thus suited for communicating a position to humans, e.g. using a position plot.

However, latitude and longitude should be used with care in mathematical expressions (including calculations in computer programs). The main reason is the singularities at the Poles, which makes longitude undefined at these points. Also near the poles the latitude/longitude grid is highly non-linear, and several errors may occur in calculations that are sufficiently accurate on other locations.

Another problematic area is the meridian at ±180° longitude, where the longitude has a discontinuity, and hence specific program code must often be written to handle this. An example of the consequences of omitting such code is the crash of the navigation systems of twelve F-22 Raptors while crossing this meridian.

n-vector

n-vector is a three parameter non-singular horizontal position representation that can replace latitude and longitude. Geometrically, it is a unit vector which is normal to the reference ellipsoid. The vector is decomposed in an Earth centered earth fixed coordinate system. It behaves the same at all Earth positions, and it holds the mathematical one-to-one property. The vector formulation makes it possible to use standard 3D vector algebra, and thus n-vector is well-suited for mathematical calculations, e.g. adding, subtracting, interpolating and averaging positions.

Using three parameters, n-vector is inconvenient for communicating a position directly to humans and before showing a position plot, a conversion to latitude/longitude might be needed.

Local Flat Earth Assumption

When carrying out several calculations within a limited area, a Cartesian coordinate system might be defined with the origin at a specified Earth-fixed position. The origin is often selected at the surface of the reference ellipsoid, with the z-axis in the vertical direction. Hence (three dimensional) position vectors relative to this coordinate frame will have two horizontal and one vertical parameter. The axes are typically selected as North-East-Down or East-North-Up, and thus this system can be viewed as a linearization of the meridians and parallels.

For small areas a local coordinate system can be convenient for relative positioning, but with increasing (horizontal) distances, errors will increase and repositioning of the tangent point may be required. The alignment along the north and east directions is not possible at the Poles, and near the Poles these directions might have significant errors (here the linearization is valid only in a very small area).

UTM

Instead of one local Cartesian grid, that needs to be repositioned as the position of interest moves, a fixed set of map projections covering the Earth can be defined. UTM is one such system, dividing the Earth into 60 longitude zones (and with UPS covering the Polar regions).

UTM is widely used, and the coordinates approximately corresponds to meters north and east. However, as a set of map-projections it has inherent distortions, and thus most calculations based on UTM will not be exact. The crossing of zones gives additional complexity.

Comparison

When deciding which parameters to use for representing position in a specific application, there are several properties that should be considered. The following table gives a summary of what to consider.

Comparison of Horizontal Position Representations		
Representation	**Pros**	**Cons**
Latitude and longitude	• Widely used • Parameters are easy to recognize by humans (well-suited for plotting)	• Singularities at the Poles • Complex behavior near the Poles • Discontinuity at the ±180° meridian
n-vector	• Nonsingular • Efficient in equations/ calculations since standard 3D vector algebra can be used • All Earth positions are treated equally	• Inconvenient for communicating a position to humans • Uses three parameters
Local Cartesian coordinate system	• Cartesian vectors in meters along the directions of north, east and down are obtained	• Can only be used for relative positioning (the tangent point must be represented by some other quantity) • Errors increase with increasing horizontal distance from the tangent point (which may require repositioning of the tangent point) • North and east directions are undefined at the Poles, and near the Poles these directions may change significantly within the area of interest

UTM	• Widely used • Approximate north and east directions • One unit corresponds approximately to one meter	• Inherent distortion (due to the map projection) gives only approximate answers for most calculations • Calculations get complex when crossing the zones • The Polar Regions are not covered

Latitude

In geography, latitude is a geographic coordinate that specifies the north–south position of a point on the Earth's surface. Latitude is an angle (defined below) which ranges from 0° at the Equator to 90° (North or South) at the poles. Lines of constant latitude, or parallels, run east–west as circles parallel to the equator. Latitude is used together with longitude to specify the precise location of features on the surface of the Earth. Without qualification the term latitude should be taken to be the geodetic latitude. Also defined are six auxiliary latitudes which are used in special applications.

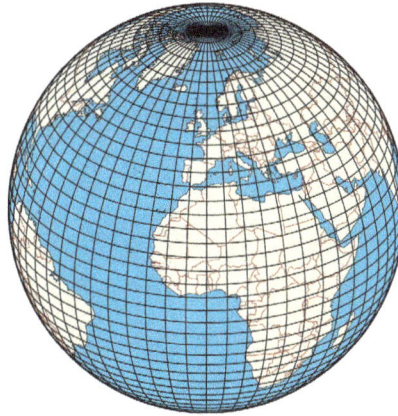

A graticule on the Earth as a sphere or an ellipsoid. The lines from pole to pole are lines of constant longitude, or meridians. The circles parallel to the equator are lines of constant latitude, or parallels. The graticule shows the latitude and longitude of points on the surface. In this example meridians are spaced at 6° intervals and parallels at 4° intervals.

Preliminaries

Two levels of abstraction are employed in the definition of latitude and longitude. In the first step the physical surface is modelled by the geoid, a surface which approximates the mean sea level over the oceans and its continuation under the land masses. The second step is to approximate the geoid by a mathematically simpler reference surface. The simplest choice for the reference surface is a sphere, but the geoid is more accurately modelled by an ellipsoid. Lines of constant latitude and longitude together constitute a graticule on the reference surface. The latitude of a point on the *actual* surface is that of the corresponding point on the reference surface, the correspondence being along the normal to the reference surface which passes through the point on the physical surface. Latitude and longitude together with some specification of height constitute a geographic coordinate system as defined in the specification of the ISO 19111 standard.

Since there are many different reference ellipsoids, the precise latitude of a feature on the surface is not unique: this is stressed in the ISO standard which states that "without the full specification of the coordinate reference system, coordinates (that is latitude and longitude) are ambiguous at best and meaningless at worst". This is of great importance in accurate applications, such as a Global Positioning System (GPS), but in common usage, where high accuracy is not required, the reference ellipsoid is not usually stated.

In English texts the latitude angle, defined below, is usually denoted by the Greek lower-case letter phi (φ or ϕ). It is measured in degrees, minutes and seconds or decimal degrees, north or south of the equator.

The precise measurement of latitude requires an understanding of the gravitational field of the Earth, either to set up theodolites or to determine GPS satellite orbits. The study of the figure of the Earth together with its gravitational field is the science of geodesy.

Here it is related to the coordinate systems of the Earth: it may be extended to cover the Moon, planets and other celestial objects by a simple change of nomenclature.

The following lists are available:

- List of cities by latitude

- List of countries by latitude

Latitude on the Sphere

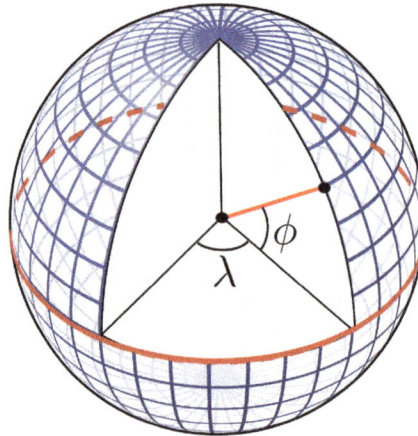

A perspective view of the Earth showing how latitude (φ) and longitude (λ) are defined on a spherical model. The graticule spacing is 10 degrees.

The Graticule on the Sphere

The graticule is formed by the lines of constant latitude and constant longitude, which are constructed with reference to the rotation axis of the Earth. The primary reference points are the poles where the axis of rotation of the Earth intersects the reference surface. Planes which contain the rotation axis intersect the surface at the meridians; and the angle between any one meridian plane and that through Greenwich (the Prime Meridian) defines the longitude: meridians are lines of constant longitude. The plane through the centre of the Earth and perpendicular to the rotation axis intersects the surface at a great circle called the Equator. Planes parallel to the equatorial plane intersect the surface in circles of constant latitude; these are the parallels. The Equator has a latitude of 0°, the North Pole has a latitude of 90° North (written 90° N or +90°), and the South Pole has a latitude of 90° South (written 90° S or −90°). The latitude of an arbitrary point is the angle between the equatorial plane and the radius to that point.

The latitude, as defined in this way for the sphere, is often termed the spherical latitude.

Named Latitudes on the Earth

The plane of the Earth's orbit about the Sun is called the ecliptic, and the plane perpendicular to the rotation axis of the Earth is the equatorial plane. The angle between the ecliptic and the equatorial plane is called variously the axial tilt, the obliquity, or the inclination of the ecliptic, and it is conventionally denoted by i. The latitude of the tropical circles is equal to i and the latitude of the polar circles is its complement ($90° - i$). The axis of rotation varies slowly over time and the values given here are those for the current epoch.

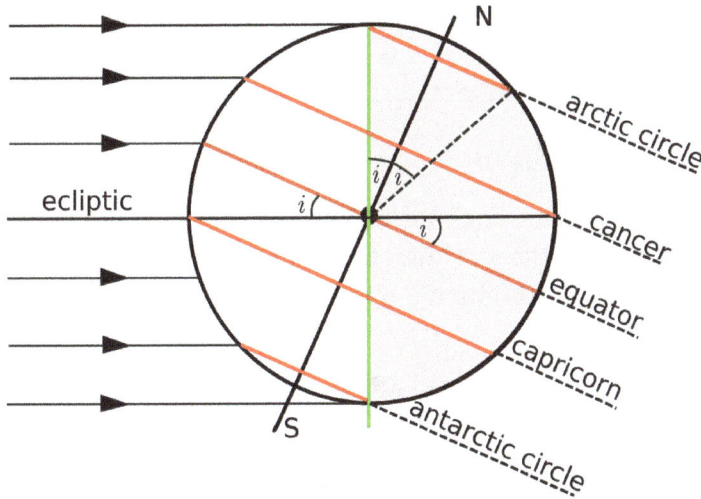

The orientation of the Earth at the December solstice.

Besides the equator, four other parallels are of significance:

Arctic Circle	66° 34′ (66.57°) N
Tropic of Cancer	23° 26′ (23.43°) N
Tropic of Capricorn	23° 26′ (23.43°) S
Antarctic Circle	66° 34′ (66.57°) S

The figure shows the geometry of a cross-section of the plane perpendicular to the ecliptic and through the centres of the Earth and the Sun at the December solstice when the Sun is overhead at some point of the Tropic of Capricorn. The south polar latitudes below the Antarctic Circle are in daylight, whilst the north polar latitudes above the Arctic Circle are in night. The situation is reversed at the June solstice, when the Sun is overhead at the Tropic of Cancer. Only at latitudes in between the two tropics is it possible for the Sun to be directly overhead (at the zenith).

On map projections there is no universal rule as to how meridians and parallels should appear. The examples below show the named parallels (as red lines) on the commonly used Mercator projection and the Transverse Mercator projection. On the former the parallels are horizontal and the meridians are vertical, whereas on the latter there is no exact relationship of parallels and meridians with horizontal and vertical: both are complicated curves.

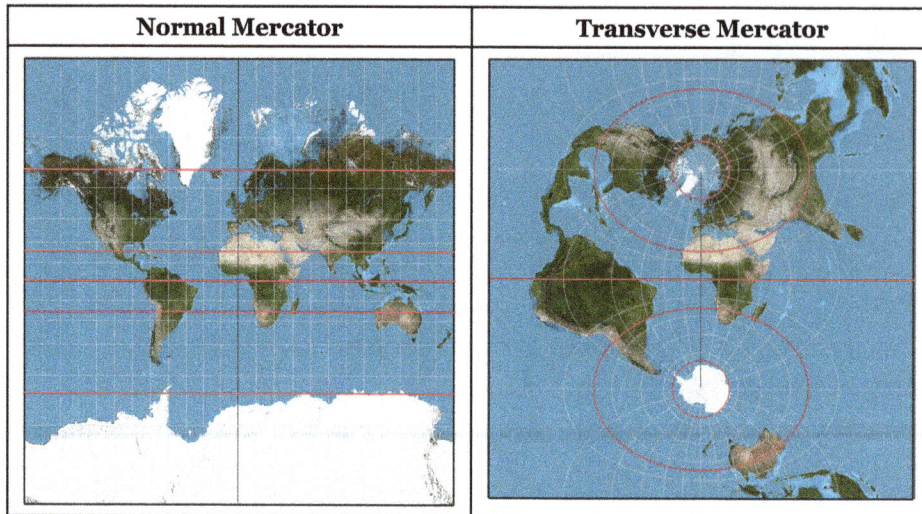

Normal Mercator	Transverse Mercator

Meridian Distance on the Sphere

On the sphere the normal passes through the centre and the latitude (φ) is therefore equal to the angle subtended at the centre by the meridian arc from the equator to the point concerned. If the meridian distance is denoted by $m(\varphi)$ then

$$m(\phi) = \frac{\pi}{180^\circ} R\phi_{degrees} = R\phi_{radians}$$

where R denotes the mean radius of the Earth. R is equal to 6,371 km or 3,959 miles. No higher accuracy is appropriate for R since higher-precision results necessitate an ellipsoid model. With this value for R the meridian length of 1 degree of latitude on the sphere is 111.2 km or 69.1 miles. The length of 1 minute of latitude is 1.853 km or 1.151 miles, used as the basis of the nautical mile.

Latitude on the Ellipsoid

Ellipsoids

In 1687 Isaac Newton published the *Philosophiæ Naturalis Principia Mathematica*, in which he proved that a rotating self-gravitating fluid body in equilibrium takes the form of an oblate ellipsoid. Newton's result was confirmed by geodetic measurements in the 18th century. An oblate ellipsoid is the three-dimensional surface generated by the rotation of an ellipse about its shorter axis (minor axis). "Oblate ellipsoid of revolution" is abbreviated to 'ellipsoid' in the remainder . (Ellipsoids which do not have an axis of symmetry are termed triaxial.)

Many different reference ellipsoids have been used in the history of geodesy. In pre-satellite days they were devised to give a good fit to the geoid over the limited area of a survey but, with the advent of GPS, it has become natural to use reference ellipsoids (such as WGS84) with centre at the centre of mass of the Earth and minor axis aligned to the rotation axis of the Earth. These geocentric ellipsoids are usually within 100 m (330 ft) of the geoid. Since latitude is defined with respect to an ellipsoid, the position of a given point is different on each ellipsoid: one cannot exactly specify the latitude and longitude of a geographical feature without specifying the ellipsoid used. Many maps maintained by national agencies are based on older ellipsoids, so one must know how the latitude and longitude values are transformed from one ellipsoid to another. GPS handsets include software to carry out datum transformations which link WGS84 to the local reference ellipsoid with its associated grid.

The Geometry of the Ellipsoid

The shape of an ellipsoid of revolution is determined by the shape of the ellipse which is rotated about its minor (shorter) axis. Two parameters are required. One is invariably the equatorial radius, which is the semi-major axis, a. The other parameter is usually (1) the polar radius or semi-minor axis, b; or (2) the (first) flattening, f; or (3) the eccentricity, e. These parameters are not independent: they are related by

$$f = \frac{a-b}{a}, \qquad e^2 = 2f - f^2, \qquad b = a(1-f) = a\sqrt{1-e^2}.$$

Many other parameters appear in the study of geodesy, geophysics and map projections but they can all be expressed in terms of one or two members of the set a, b, f and e. Both f and e are small and often appear in series expansions in calculations; they are of the order $1/300$ and 0.08 respectively. Values for a number of ellipsoids are given in Figure of the Earth. Reference ellipsoids are usually defined by the semi-major axis and the *inverse* flattening, $1/f$. For example, the defining values for the WGS84 ellipsoid, used by all GPS devices, are

- a (equatorial radius): 6378137.0 m exactly
- $\frac{1}{f}$ (inverse flattening): 298.257223563 exactly

from which are derived

- b (polar radius): 6356752.3142 m
- e^2 (eccentricity squared): 0.00669437999014

The difference between the semi-major and semi-minor axes is about 21 km (13 miles) and as fraction of the semi-major axis it equals the flattening; on a computer the ellipsoid could be sized as 300 by 299 pixels. This would barely be distinguishable from a 300-by-300-pixel sphere, so illustrations usually exaggerate the flattening.

Geodetic and Geocentric Latitudes

The graticule on the ellipsoid is constructed in exactly the same way as on the sphere. The normal at a point on the surface of an ellipsoid does not pass through the centre, except for points on the equator or at the poles, but the definition of latitude remains unchanged as the angle between the normal and the equatorial plane. The terminology for latitude must be made more precise by distinguishing:

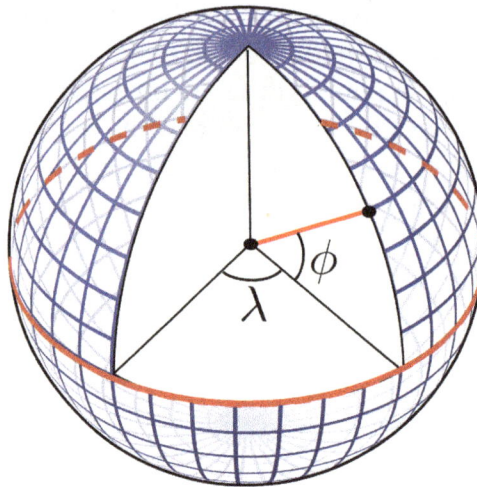

The definition of geodetic latitude (φ) and longitude (λ) on an ellipsoid. The normal to the surface does not pass through the centre, except at the equator and at the poles.

- Geodetic latitude: the angle between the normal and the equatorial plane. The standard notation in English publications is φ. This is the definition assumed when the word latitude is used without qualification. The definition must be accompanied with a specification of the ellipsoid.

- Geocentric latitude: the angle between the radius (from centre to the point on the surface) and the equatorial plane. (Figure below). There is no standard notation: examples from various texts include ψ, q, φ', φ_c, φ_g. Here we use ψ.

- Spherical latitude: the angle between the normal to a spherical reference surface and the equatorial plane.

- Geographic latitude must be used with care. Some authors use it as a synonym for geodetic latitude whilst others use it as an alternative to the astronomical latitude.

- Latitude (unqualified) should normally refer to the geodetic latitude.

The importance of specifying the reference datum may be illustrated by a simple example. On the reference ellipsoid for WGS84, the centre of the Eiffel Tower has a geodetic latitude of 48° 51′ 29″ N, or 48.8583° N and longitude of 2° 17′ 40″ E or 2.2944°E. The same coordinates on the datum ED50 define a point on the ground which is 140 metres (460 feet) distant from the tower. A web search may produce several different values for the latitude of the tower; the reference ellipsoid is rarely specified.

Length of a Degree of Latitude

In Meridian arc and standard texts it is shown that the distance along a meridian from latitude φ to the equator is given by (φ in radians)

$$m(\phi) = \int_0^\phi M(\phi')d\phi' = a(1-e^2)\int_0^\phi \left(1-e^2\sin^2\phi'\right)^{-\frac{3}{2}}d\phi'$$

where $M(\varphi)$ is the meridional radius of curvature.

The distance from the equator to the pole is

$$m_p = m\left(\frac{\pi}{2}\right)$$

For WGS84 this distance is 10001.965729 km.

The evaluation of the meridian distance integral is central to many studies in geodesy and map projection. It can be evaluated by expanding the integral by the binomial series and integrating term by term. The length of the meridian arc between two given latitudes is given by replacing the limits of the integral by the latitudes concerned. The length of a *small* meridian arc is given by

$$\delta m(\phi) = M(\phi)\delta\phi = a(1-e^2)\left(1-e^2\sin^2\phi\right)^{-\frac{3}{2}}\delta\phi$$

ϕ	Δ^1_{lat}	Δ^1_{long}
0°	110.574 km	111.320 km
15°	110.649 km	107.550 km
30°	110.852 km	96.486 km
45°	111.132 km	78.847 km
60°	111.412 km	55.800 km
75°	111.618 km	28.902 km
90°	111.694 km	0.000 km

When the latitude difference is 1 degree, corresponding to $\dfrac{\pi}{180}$ radians, the arc distance is about

$$\Delta^1_{lat} = \frac{\pi a\left(1-e^2\right)}{180°\left(1-e^2\sin^2\phi\right)^{\frac{3}{2}}}$$

The distance in metres (correct to 0.01 metre) between latitudes ϕ − 0.5 degrees and ϕ + 0.5 degrees on the WGS84 spheroid is

$$\Delta^1_{lat} = 111132.954 - 559.822\cos 2\phi + 1.175\cos 4\phi$$

The variation of this distance with latitude (on WGS84) is shown in the table along with the length of a degree of longitude (east-west distance):

$$\Delta^1_{long} = \frac{\pi a\cos\phi}{180°\sqrt{1-e^2\sin^2\phi}}$$

A calculator for any latitude is provided by the U.S. Government's National Geospatial-Intelligence Agency (NGA).

Historically a nautical mile was defined as the length of one minute of arc along a meridian of a spherical earth. An ellipsoid model leads to a variation of the nautical mile with latitude. This was resolved by defining the nautical mile to be exactly 1,852 metres.

Auxiliary Latitudes

There are six auxiliary latitudes that have applications to special problems in geodesy, geophysics and the theory of map projections:

- Geocentric latitude

- Reduced (or parametric) latitude

- Rectifying latitude

- Authalic latitude

- Conformal latitude

- Isometric latitude

The definitions given all relate to locations on the reference ellipsoid but the first two auxiliary latitudes, like the geodetic latitude, can be extended to define a three-dimensional geographic co-ordinate system as discussed below. The remaining latitudes are not used in this way; they are used *only* as intermediate constructs in map projections of the reference ellipsoid to the plane or in calculations of geodesics on the ellipsoid. Their numerical values are not of interest. For example, no one would need to calculate the authalic latitude of the Eiffel Tower.

The expressions below give the auxiliary latitudes in terms of the geodetic latitude, the semi-major axis, a, and the eccentricity, e. The forms given are, apart from notational variants, those in the standard reference for map projections, namely "Map projections: a working manual" by J. P. Snyder. Derivations of these expressions may be found in Adams and online publications by Osborne and Rapp.

Geocentric Latitude

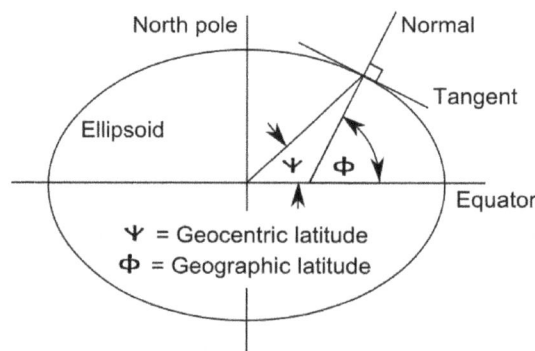

The definition of geodetic (or geographic) and geocentric latitudes.

The geocentric latitude is the angle between the equatorial plane and the radius from the centre to a point on the surface. The relation between the geocentric latitude (ψ) and the geodetic latitude

(φ) is derived in the above references as

$$\psi(\phi) = \tan^{-1}\left((1-e^2)\tan\phi\right).$$

The geodetic and geocentric latitudes are equal at the equator and at the poles but at other latitudes they differ by a few minutes of arc. Taking the value of the squared eccentricity as 0.0067 (it depends on the choice of ellipsoid) the maximum difference of $\phi - \psi$ may be shown to be about 11.5 minutes of arc at a geodetic latitude of approximately 45° 6'.

Reduced (or Parametric) Latitude

The reduced or parametric latitude, β, is defined by the radius drawn from the centre of the ellipsoid to that point Q on the surrounding sphere (of radius a) which is the projection parallel to the Earth's axis of a point P on the ellipsoid at latitude φ. It was introduced by Legendre and Bessel who solved problems for geodesics on the ellipsoid by transforming them to an equivalent problem for spherical geodesics by using this smaller latitude. Bessel's notation, $u(\varphi)$, is also used in the current literature. The reduced latitude is related to the geodetic latitude by:

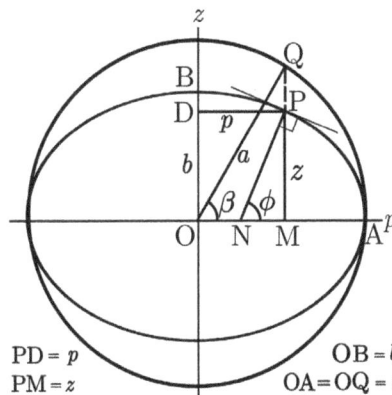

Definition of the reduced latitude (β) on the ellipsoid.

$$\beta(\phi) = \tan^{-1}\left(\sqrt{1-e^2}\,\tan\phi\right)$$

The alternative name arises from the parameterization of the equation of the ellipse describing a meridian section. In terms of Cartesian coordinates p, the distance from the minor axis, and z, the distance above the equatorial plane, the equation of the ellipse is:

$$\frac{p^2}{a^2} + \frac{z^2}{b^2} = 1.$$

The Cartesian coordinates of the point are parameterized by

$$p = a\cos\beta, \qquad z = b\sin\beta;$$

Cayley suggested the term *parametric latitude* because of the form of these equations.

The reduced latitude is not used in the theory of map projections. Its most important application is in the theory of ellipsoid geodesics. (Vincenty, Karney).

Rectifying Latitude

The rectifying latitude, μ, is the meridian distance scaled so that its value at the poles is equal to 90 degrees or $\dfrac{\pi}{2}$ radians:

$$\mu(\phi) = \frac{\pi}{2} \frac{m(\phi)}{m_p}$$

where the meridian distance from the equator to a latitude φ is

$$m(\phi) = a(1-e^2)\int_0^\phi \left(1-e^2 \sin^2 \phi'\right)^{-\frac{3}{2}} d\phi',$$

and the length of the meridian quadrant from the equator to the pole (the polar distance) is

$$m_p = m\left(\frac{\pi}{2}\right).$$

Using the rectifying latitude to define a latitude on a sphere of radius

$$R = \frac{2m_p}{\pi}$$

defines a projection from the ellipsoid to the sphere such that all meridians have true length and uniform scale. The sphere may then be projected to the plane with an equirectangular projection to give a double projection from the ellipsoid to the plane such that all meridians have true length and uniform meridian scale. An example of the use of the rectifying latitude is the Equidistant conic projection. (Snyder, Section 16). The rectifying latitude is also of great importance in the construction of the Transverse Mercator projection.

Authalic Latitude

The authalic (Greek for same area) latitude, ξ, gives an area-preserving transformation to a sphere.

$$\xi(\phi) = \sin^{-1}\left(\frac{q(\phi)}{q_p}\right)$$

where

$$
\begin{aligned}
q(\phi) &= \frac{(1-e^2)\sin\phi}{1-e^2\sin^2\phi} - \frac{1-e^2}{2e}\ln\left(\frac{1-e\sin\phi}{1+e\sin\phi}\right) \\
&= \frac{(1-e^2)\sin\phi}{1-e^2\sin^2\phi} + \frac{1-e^2}{e}\tanh^{-1}(e\sin\phi)
\end{aligned}
$$

and

$$q_p = q\left(\frac{\pi}{2}\right) = 1 - \frac{1-e^2}{2e}\ln\left(\frac{1-e}{1+e}\right) = 1 + \frac{1-e^2}{e}\tanh^{-1}e$$

and the radius of the sphere is taken as

$$R_q = a\sqrt{\frac{q_p}{2}}.$$

An example of the use of the authalic latitude is the Albers equal-area conic projection.

Conformal Latitude

The conformal latitude, χ, gives an angle-preserving (conformal) transformation to the sphere.

$$\chi(\phi) = 2\tan^{-1}\left[\left(\frac{1+\sin\phi}{1-\sin\phi}\right)\left(\frac{1-e\sin\phi}{1+e\sin\phi}\right)^e\right]^{\frac{1}{2}} - \frac{\pi}{2}$$

$$2\tan^{-1}\left[\tan\left(\frac{\phi}{2}+\frac{\pi}{4}\right)\left(\frac{1-e\sin\phi}{1+e\sin\phi}\right)^{\frac{e}{2}}\right] - \frac{\pi}{2}$$

$$= \sin^{-1}\left[\tanh\left(\tanh^{-1}(\sin\phi) - e\tanh^{-1}(e\sin\phi)\right)\right]$$

$$= gd\left[gd^{-1}(\phi) - e\tanh^{-1}(e\sin\phi)\right]$$

where gd(x) is the Gudermannian function. The conformal latitude defines a transformation from the ellipsoid to a sphere of *arbitrary* radius such that the angle of intersection between any two lines on the ellipsoid is the same as the corresponding angle on the sphere (so that the shape of *small* elements is well preserved). A further conformal transformation from the sphere to the plane gives a conformal double projection from the ellipsoid to the plane. This is not the only way of generating such a conformal projection. For example, the 'exact' version of the Transverse Mercator projection on the ellipsoid is not a double projection. (It does, however, involve a generalisation of the conformal latitude to the complex plane).

Isometric Latitude

The isometric latitude is conventionally denoted by ψ (different than geocentric latitude): it is used in the development of the ellipsoidal versions of the normal Mercator projection and the Transverse Mercator projection. The name "isometric" arises from the fact that at any point on the ellipsoid equal increments of ψ and longitude λ give rise to equal distance displacements along the meridians and parallels respectively. The graticule defined by the lines of constant ψ and constant λ, divides the surface of the ellipsoid into a mesh of squares (of varying size). The isometric latitude is zero at the equator but rapidly diverges from the geodetic latitude, tending to infinity at the poles. The conventional notation is given in Snyder:

$$\psi(\phi) = \ln\left[\tan\left(\frac{\pi}{4}+\frac{\phi}{2}\right)\right] + \frac{e}{2}\ln\left[\frac{1-e\sin\phi}{1+e\sin\phi}\right]$$
$$= \tanh^{-1}(\sin\phi) - e\tanh^{-1}(e\sin\phi)$$
$$= \mathrm{gd}^{-1}(\phi) - e\tanh^{-1}(e\sin\phi).$$

For the *normal* Mercator projection (on the ellipsoid) this function defines the spacing of the parallels: if the length of the equator on the projection is E (units of length or pixels) then the distance, y, of a parallel of latitude φ from the equator is

$$y(\phi) = \frac{E}{2\pi}\psi(\phi).$$

The isometric latitude ψ is closely related to the conformal latitude χ:

$$\psi(\phi) = \mathrm{gd}^{-1}\chi(\phi).$$

Inverse Formulae and Series

The expressions for the geocentric and reduced latitudes may be inverted directly but this is impossible in the four remaining cases: the rectifying, authalic, conformal, and isometric latitudes. There are two methods of proceeding. The first is a numerical inversion of the defining equation for each and every particular value of the auxiliary latitude. The methods available are fixed-point iteration and Newton–Raphson root finding. The other, more useful, approach is to express the auxiliary latitude as a series in terms of the geodetic latitude and then invert the series by the method of Lagrange reversion. Such series are presented by Adams who uses Taylor series expansions and gives coefficients in terms of the eccentricity. Osborne derives series to arbitrary order by using the computer algebra package Maxima and expresses the coefficients in terms of both eccentricity and flattening. The series method is not applicable to the isometric latitude and one must use the conformal latitude in an intermediate step.

Numerical Comparison of Auxiliary Latitudes

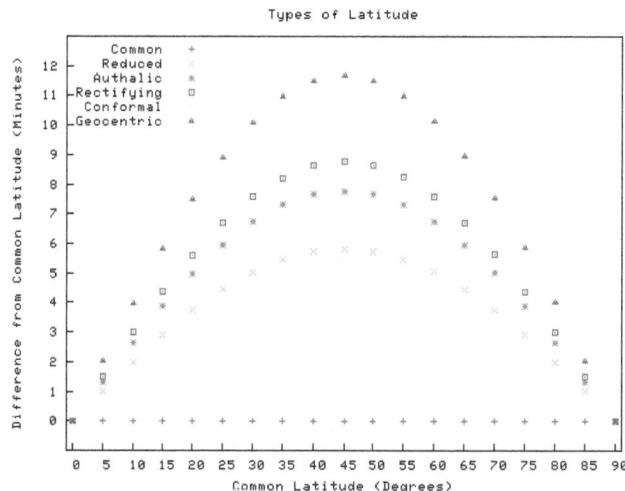

The following plot shows the magnitude of the difference between the geodetic latitude, (denoted as the "common" latitude on the plot), and the auxiliary latitudes other than the isometric latitude (which diverges to infinity at the poles). In every case the geodetic latitude is the greater. The differences shown on the plot are in arc minutes. The horizontal resolution of the plot fails to make clear that the maxima of the curves are not at 45° but calculation shows that they are within a few arc minutes of 45°. Some representative data points are given in the table following the plot. Note the closeness of the conformal and geocentric latitudes. This was exploited in the days of hand calculators to expedite the construction of map projections.

φ	Approximate difference from geodetic latitude (φ)				
	Reduced $\varphi - \beta$	Authalic $\varphi - \xi$	Rectifying $\varphi - \mu$	Conformal $\varphi - \chi$	Geocentric $\varphi - \psi$
0°	0.00'	0.00'	0.00'	0.00'	0.00'
15°	2.91'	3.89'	4.37'	5.82'	5.82'
30°	5.05'	6.73'	7.57'	10.09'	10.09'
45°	5.84'	7.78'	8.76'	11.67'	11.67'
60°	5.06'	6.75'	7.59'	10.12'	10.13'
75°	2.92'	3.90'	4.39'	5.85'	5.85'
90°	0.00'	0.00'	0.00'	0.00'	0.00'

Latitude and Coordinate Systems

The geodetic latitude, or any of the auxiliary latitudes defined on the reference ellipsoid, constitutes with longitude a two-dimensional coordinate system on that ellipsoid. To define the position of an arbitrary point it is necessary to extend such a coordinate system into three dimensions. Three latitudes are used in this way: the geodetic, geocentric and reduced latitudes are used in geodetic coordinates, spherical polar coordinates and ellipsoidal coordinates respectively.

Geodetic Coordinates

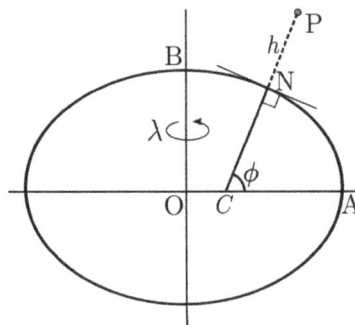

Geodetic coordinates P(ϕ,λ,h)

At an arbitrary point P consider the line PN which is normal to the reference ellipsoid. The geodetic coordinates P(ϕ,λ,h) are the latitude and longitude of the point N on the ellipsoid and the distance PN. This height differs from the height above the geoid or a reference height such as that above mean sea level at a specified location. The direction of PN will also differ from the direction of a vertical plumb line. The relation of these different heights requires knowledge of the shape of the geoid and also the gravity field of the Earth.

Spherical Polar Coordinates

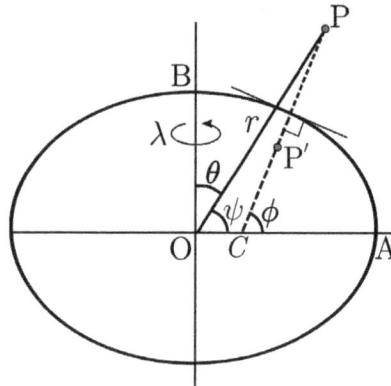

Geocentric coordinate related to spherical polar coordinates $P(r,\theta,\lambda)$

The geocentric latitude ψ is the complement of the polar angle θ in conventional spherical polar coordinates in which the coordinates of a point are $P(r,\theta,\lambda)$ where r is the distance of P from the centre O, θ is the angle between the radius vector and the polar axis and λ is longitude. Since the normal at a general point on the ellipsoid does not pass through the centre it is clear that points on the normal, which all have the same geodetic latitude, will have differing geocentric latitudes. Spherical polar coordinate systems are used in the analysis of the gravity field.

Ellipsoidal Coordinates

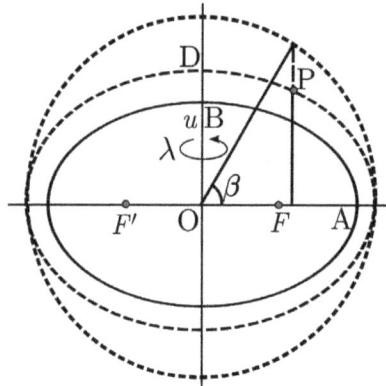

Ellipsoidal coordinates $P(u,\beta,\lambda)$

The reduced latitude can also be extended to a three-dimensional coordinate system. For a point P not on the reference ellipsoid (semi-axes OA and OB) construct an auxiliary ellipsoid which is confocal (same foci F, F′) with the reference ellipsoid: the necessary condition is that the product ae of semi-major axis and eccentricity is the same for both ellipsoids. Let u be the semi-minor axis (OD) of the auxiliary ellipsoid. Further let β be the reduced latitude of P on the auxiliary ellipsoid. The set (u,β,λ) define the ellipsoid coordinates. These coordinates are the natural choice in models of the gravity field for a uniform distribution of mass bounded by the reference ellipsoid.

Coordinate Conversions

The relations between the above coordinate systems, and also Cartesian coordinates are not presented here. The transformation between geodetic and Cartesian coordinates may be found in Geographic coordinate conversion. The relation of Cartesian and spherical polars is given in Spherical coordinate system. The relation of Cartesian and ellipsoidal coordinates is discussed in Torge.

Astronomical Latitude

Astronomical latitude (Φ) is the angle between the equatorial plane and the true vertical at a point on the surface. The true vertical, the direction of a plumb line, is also the direction of the gravity acceleration, the resultant of the gravitational acceleration (mass-based) and the centrifugal acceleration at that latitude. Astronomic latitude is calculated from angles measured between the zenith and stars whose declination is accurately known.

In general the true vertical at a point on the surface does not exactly coincide with either the normal to the reference ellipsoid or the normal to the geoid. The angle between the astronomic and geodetic normals is usually a few seconds of arc but it is important in geodesy. The reason why it differs from the normal to the geoid is, because the geoid is an idealized, theoretical shape "at mean sea level". Points on the real surface of the earth are usually above or below this idealized geoid surface and here the true vertical can vary slightly. Also, the true vertical at a point at a specific time is influenced by tidal forces, which the theoretical geoid averages out.

Astronomical latitude is different than declination, the coordinate astronomers use in a similar way to specify the angular position of stars north/south of the celestial equator. nor with ecliptic latitude, the coordinate that astronomers use to specify the angular position of stars north/south of the ecliptic.

Longitude

Longitude, Australian and British also is a geographic coordinate that specifies the east-west position of a point on the Earth's surface. It is an angular measurement, usually expressed in degrees and denoted by the Greek letter lambda (λ). Meridians (lines running from the North Pole to the South Pole) connect points with the same longitude. By convention, one of these, the Prime Meridian, which passes through the Royal Observatory, Greenwich, England, was allocated the position of zero degrees longitude. The longitude of other places is measured as the angle east or west from the Prime Meridian, ranging from 0° at the Prime Meridian to +180° eastward and −180° westward. Specifically, it is the angle between a plane containing the Prime Meridian and a plane containing the North Pole, South Pole and the location in question. (This forms a right-handed coordinate system with the z axis (right hand thumb) pointing from the Earth's center toward the North Pole and the x axis (right hand index finger) extending from Earth's center through the equator at the Prime Meridian.)

A location's north–south position along a meridian is given by its latitude, which is approximately the angle between the local vertical and the plane of the Equator.

If the Earth were perfectly spherical and homogeneous, then the longitude at a point would be equal to the angle between a vertical north–south plane through that point and the plane of the Greenwich meridian. Everywhere on Earth the vertical north–south plane would contain the Earth's axis. But the Earth is not homogeneous, and has mountains—which have gravity and so can shift the vertical plane away from the Earth's axis. The vertical north–south plane still intersects the plane of the Greenwich meridian at some angle; that angle is the astronomical longitude, calculated from star observations. The longitude shown on maps and GPS devices is the angle between the Greenwich plane and a not-quite-vertical plane through the point; the not-quite-vertical plane is perpendicular to the surface of the spheroid chosen to approximate the Earth's sea-level surface, rather than perpendicular to the sea-level surface itself.

History

The measurement of longitude is important both to cartography and for ocean navigation. Mariners and explorers for most of history struggled to determine longitude. Finding a method of determining longitude took centuries, resulting in the history of longitude recording the effort of some of the greatest scientific minds.

Latitude was calculated by observing with quadrant or astrolabe the altitude of the sun or of charted stars above the horizon, but longitude is harder.

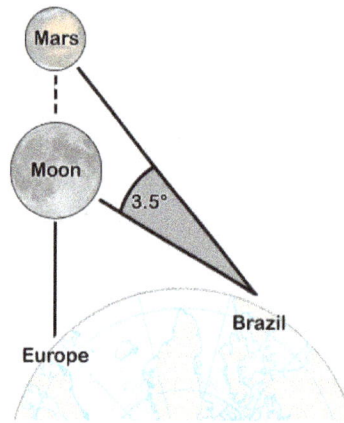

Amerigo Vespucci's means of determining longitude

Amerigo Vespucci was perhaps the first European to proffer a solution, after devoting a great deal of time and energy studying the problem during his sojourns in the New World:

As to longitude, I declare that I found so much difficulty in determining it that I was put to great pains to ascertain the east-west distance I had covered. The final result of my labours was that I found nothing better to do than to watch for and take observations at night of the conjunction of one planet with another, and especially of the conjunction of the moon with the other planets, because the moon is swifter in her course than any other planet. I compared my observations with an almanac. After I had made experiments many nights, one night, the twenty-third of August 1499, there was a conjunction of the moon with Mars, which according to the almanac was to occur at midnight or a half hour before. I found that...at midnight Mars's position was three and a half degrees to the east.

By comparing the positions of the moon and Mars with their anticipated positions, Vespucci was able to crudely deduce his longitude. But this method had several limitations: First, it required the occurrence of a specific astronomical event (in this case, Mars passing through the same right ascension as the moon), and the observer needed to anticipate this event via an astronomical almanac. One needed also to know the precise time, which was difficult to ascertain in foreign lands.

Finally, it required a stable viewing platform, rendering the technique useless on the rolling deck of a ship at sea.

John Harrison solved the greatest problem of his day.

In 1612 Galileo Galilei demonstrated that with sufficiently accurate knowledge of the orbits of the moons of Jupiter one could use their positions as a universal clock and this would make possible the determination of longitude, but the method he devised was impracticable for navigators on ships because of their instability. In 1714 the British government passed the Longitude Act which offered large financial rewards to the first person to demonstrate a practical method for determining the longitude of a ship at sea. These rewards motivated many to search for a solution.

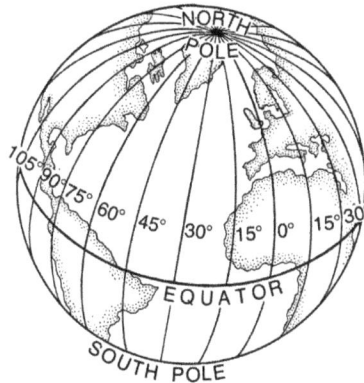

Drawing of Earth with longitudes but without latitudes.

John Harrison, a self-educated English clockmaker, invented the marine chronometer, the key piece in solving the problem of accurately establishing longitude at sea, thus revolutionising and extending the possibility of safe long distance sea travel. Though the Board of Longitude rewarded John Harrison for his marine chronometer in 1773, chronometers remained very expensive and the lunar distance method continued to be used for decades. Finally, the combination of the availability of marine chronometers and wireless telegraph time signals put an end to the use of lunars in the 20th century.

Unlike latitude, which has the equator as a natural starting position, there is no natural starting position for longitude. Therefore, a reference meridian had to be chosen. It was a popular practice to use a nation's capital as the starting point, but other locations were also used. While British cartographers had long used the Greenwich meridian in London, other references were used elsewhere, including El Hierro, Rome, Copenhagen, Jerusalem, Saint Petersburg, Pisa, Paris, Philadelphia, and Washington D.C. In 1884 the International Meridian Conference adopted the Greenwich meridian as the *universal Prime Meridian* or *zero point of longitude*.

Noting and Calculating Longitude

Longitude is given as an angular measurement ranging from 0° at the Prime Meridian to +180° eastward and –180° westward. The Greek letter λ (lambda), is used to denote the location of a place on Earth east or west of the Prime Meridian.

Each degree of longitude is sub-divided into 60 minutes, each of which is divided into 60 seconds. A longitude is thus specified in sexagesimal notation as 23° 27′ 30″ E. For higher precision, the seconds are specified with a decimal fraction. An alternative representation uses degrees and minutes, where parts of a minute are expressed in decimal notation with a fraction, thus: 23° 27.5′ E. Degrees may also be expressed as a decimal fraction: 23.45833° E. For calculations, the angular measure may be converted to radians, so longitude may also be expressed in this manner as a signed fraction of π (pi), or an unsigned fraction of 2π.

For calculations, the West/East suffix is replaced by a negative sign in the western hemisphere. Confusingly, the convention of negative for East is also sometimes seen. The preferred convention—that East is positive—is consistent with a right-handed Cartesian coordinate system, with the North Pole up. A specific longitude may then be combined with a specific latitude (usually positive in the northern hemisphere) to give a precise position on the Earth's surface.

There is no other physical principle determining longitude directly but with time. Longitude at a point may be determined by calculating the time difference between that at its location and Coordinated Universal Time (UTC). Since there are 24 hours in a day and 360 degrees in a circle, the sun moves across the sky at a rate of 15 degrees per hour (360° ÷ 24 hours = 15° per hour). So if the time zone a person is in is three hours ahead of UTC then that person is near 45° longitude (3 hours × 15° per hour = 45°). The word *near* is used because the point might not be at the center of the time zone; also the time zones are defined politically, so their centers and boundaries often do not lie on meridians at multiples of 15°. In order to perform this calculation, however, a person needs to have a chronometer (watch) set to UTC and needs to determine local time by solar or astronomical observation.

Singularity and Discontinuity of Longitude

Note that the longitude is singular at the Poles and calculations that are sufficiently accurate for other positions, may be inaccurate at or near the Poles. Also the discontinuity at the ±180° meridian must be handled with care in calculations. An example is a calculation of east displacement by subtracting two longitudes, which gives the wrong answer if the two positions are on either side of this meridian. To avoid these complexities, consider replacing latitude and longitude with another horizontal position representation in calculation.

Plate Movement and Longitude

The Earth's tectonic plates move relative to one another in different directions at speeds on the order of 50 to 100mm per year. So points on the Earth's surface on different plates are always in motion relative to one another, for example, the longitudinal difference between a point on the Equator in Uganda, on the African Plate, and a point on the Equator in Ecuador, on the South American Plate, is increasing by about 0.0014 arcseconds per year. These tectonic movements likewise affect latitude.

If a global reference frame (such as WGS84, for example) is used, the longitude of a place on the surface will change from year to year. To minimize this change, when dealing just with points on a single plate, a different reference frame can be used, whose coordinates are fixed to a particular plate, such as "NAD83" for North America or "ETRS89" for Europe.

Length of a Degree of Longitude

The length of a degree of longitude (east-west distance) depends only on the radius of a circle of latitude. For a sphere of radius a that radius at latitude φ is $a \cos \varphi$, and the length of a one-degree (or $\dfrac{\pi}{180}$ radian) arc along a circle of latitude is

$$\Delta^1_{long} = \frac{\pi}{180^\circ} a \cos \phi$$

φ	Δ^1_{Lat}	Δ^1_{Long}
0°	110.574 km	111.320 km
15°	110.649 km	107.551 km
30°	110.852 km	96.486 km
45°	111.132 km	78.847 km
60°	111.412 km	55.800 km
75°	111.618 km	28.902 km
90°	111.694 km	0.000 km

When the Earth is modelled by an ellipsoid this arc length becomes

$$\Delta^1_{long} = \frac{\pi a \cos \phi}{180^\circ \sqrt{1 - e^2 \sin^2 \phi}}$$

where e, the eccentricity of the ellipsoid, is related to the major and minor axes (the equatorial and polar radii respectively) by

$$e^2 = \frac{a^2 - b^2}{a^2}$$

An alternative formula is

$$\Delta^1_{long} = \frac{\pi}{180^\circ} a \cos \psi \quad \text{where } \tan \psi = \frac{b}{a} \tan \phi$$

Cos φ decreases from 1 at the equator to 0 at the poles, which measures how circles of latitude shrink from the equator to a point at the pole, so the length of a degree of longitude decreases likewise. This contrasts with the small (1%) increase in the length of a degree of latitude (north-south distance), equator to pole. The table shows both for the WGS84 ellipsoid with a = 6378137.0 m and b = 6356752.3142 m. Note that the distance between two points 1 degree apart on the same circle of latitude, measured along that circle of latitude, is slightly more than the shortest (geodesic) distance between those points (unless on the equator, where these are equal); the difference is less than 0.6 m (2 ft).

A geographical mile is defined to be the length of one minute of arc along the equator (one equatorial minute of longitude), so a degree of longitude along the equator is exactly 60 geographical miles, as there are 60 minutes in a degree.

Longitude on Bodies Other than Earth

Planetary co-ordinate systems are defined relative to their mean axis of rotation and various definitions of longitude depending on the body. The longitude systems of most of those bodies with observable rigid surfaces have been defined by references to a surface feature such as a crater. The north pole is that pole of rotation that lies on the north side of the invariable plane of the solar system (near the ecliptic). The location of the Prime Meridian as well as the position of body's north pole on the celestial sphere may vary with time due to precession of the axis of rotation of the planet (or satellite). If the position angle of the body's Prime Meridian increases with time, the body has a direct (or prograde) rotation; otherwise the rotation is said to be retrograde.

In the absence of other information, the axis of rotation is assumed to be normal to the mean orbital plane; Mercury and most of the satellites are in this category. For many of the satellites, it is assumed that the rotation rate is equal to the mean orbital period. In the case of the giant planets, since their surface features are constantly changing and moving at various rates, the rotation of their magnetic fields is used as a reference instead. In the case of the Sun, even this criterion fails (because its magnetosphere is very complex and does not really rotate in a steady fashion), and an agreed-upon value for the rotation of its equator is used instead.

For *planetographic longitude*, west longitudes (i.e., longitudes measured positively to the west) are used when the rotation is prograde, and east longitudes (i.e., longitudes measured positively to the east) when the rotation is retrograde. In simpler terms, imagine a distant, non-orbiting observer viewing a planet as it rotates. Also suppose that this observer is within the plane of the planet's equator. A point on the Equator that passes directly in front of this observer later in time has a higher planetographic longitude than a point that did so earlier in time.

However, *planetocentric longitude* is always measured positively to the east, regardless of which way the planet rotates. *East* is defined as the counter-clockwise direction around the planet, as seen from above its north pole, and the north pole is whichever pole more closely aligns with the Earth's north pole. Longitudes traditionally have been written using "E" or "W" instead of "+" or "−" to indicate this polarity. For example, the following all mean the same thing:

- −91°

- 91°W

- +269°

- 269°E.

The reference surfaces for some planets (such as Earth and Mars) are ellipsoids of revolution for which the equatorial radius is larger than the polar radius; in other words, they are oblate spheroids. Smaller bodies (Io, Mimas, etc.) tend to be better approximated by triaxial ellipsoids; however, triaxial ellipsoids would render many computations more complicated, especially those related to map projections. Many projections would lose their elegant and popular properties. For this reason spherical reference surfaces are frequently used in mapping programs.

The modern standard for maps of Mars (since about 2002) is to use planetocentric coordinates. The meridian of Mars is located at Airy-0 crater.

Tidally-locked bodies have a natural reference longitude passing through the point nearest to their parent body: 0° the center of the primary-facing hemisphere, 90° the center of the leading hemisphere, 180° the center of the anti-primary hemisphere, and 270° the center of the trailing hemisphere. However, libration due to non-circular orbits or axial tilts causes this point to move around any fixed point on the celestial body like an analemma.

Figure of the Earth

The expression figure of the Earth has various meanings in geodesy according to the way it is used and the precision with which the Earth's size and shape is to be defined. While the sphere is a close approximation of the true figure of the Earth and satisfactory for many purposes, geodesists have developed several models that more closely approximate the shape of the Earth so that coordinate systems can serve the precise needs of navigation, surveying, cadastre, land use, and various other concerns.

Need for Models of the Figure of the Earth

The actual topographic surface is most apparent with its variety of land forms and water areas. This is, in fact, the surface on which actual Earth measurements are made. However, it is not feasible for exact mathematical analysis, because the formulas which would be required to take the irregularities into account would necessitate a prohibitive amount of computation. The topographic surface is generally the concern of topographers and hydrographers.

The Pythagorean concept of a spherical Earth offers a simple surface that is mathematically easy to deal with. Many astronomical and navigational computations use it as a surface representing the Earth. While the sphere is a close approximation of the true figure of the Earth and satisfactory for many purposes, to the geodesists interested in the measurement of long distances on the scale of continents and oceans, a more exact figure is necessary. Closer approximations range from modelling the shape of the surface of the entire Earth as an oblate spheroid or an oblate ellipsoid, to the use of spherical harmonics or local approximations in terms of local reference ellipsoids.

The idea of a planar or flat surface for Earth, however, is still sufficient for surveys of small areas, as the local topography is far more significant than the curvature. Plane-table surveys are made for relatively small areas, and no account is taken of the curvature of the Earth. A survey of a city would likely be computed as though the Earth were a plane surface the size of the city. For such small areas, exact positions can be determined relative to each other without considering the size and shape of the entire Earth.

In the mid- to late 20th century, research across the geosciences contributed to drastic improvements in the accuracy of the figure of the Earth. The primary utility (and the motivation for funding, mainly from the military) of this improved accuracy was to provide geographical and gravitational data for the inertial guidance systems of ballistic missiles. This funding also drove the expansion of geoscientific disciplines, fostering the creation and growth of various geoscience departments at many universities.

Models of the Figure of the Earth

The models for the figure of the Earth vary in the way they are used, in their complexity, and in the accuracy with which they represent the size and shape of the Earth.

Sphere

The simplest model for the shape of the entire Earth is a sphere. The Earth's radius is the distance from Earth's center to its surface, about 6,371 kilometers (3,959 mi). While "radius" normally is a characteristic of perfect spheres, the Earth deviates from a perfect sphere by only a third of a percent, sufficiently close to treat it as a sphere in many contexts and justifying the term "the radius of the Earth".

A view across a 20-km-wide bay in the coast of Spain. Note the curvature of the Earth hiding the base of the buildings on the far shore.

The concept of a spherical Earth dates back to around the 6th century BC, but remained a matter of philosophical speculation until the 3rd century BC. The first scientific estimation of the radius of the earth was given by Eratosthenes about 240 BC, with estimates of the accuracy of Eratosthenes's measurement ranging from 2% to 15%.

The Earth is only approximately spherical, so no single value serves as its natural radius. Distances from points on the surface to the center range from 6,353 km to 6,384 km (3,947 – 3,968 mi). Several different ways of modeling the Earth as a sphere each yield a mean radius of 6,371 kilometers (3,959 mi). Regardless of the model, any radius falls between the polar minimum of about 6,357 km and the equatorial maximum of about 6,378 km (3,950 – 3,963 mi). The difference 21 kilometers (13 mi) correspond to the polar radius being approximately 0.3% shorter than the equator radius.

Ellipsoid of Revolution

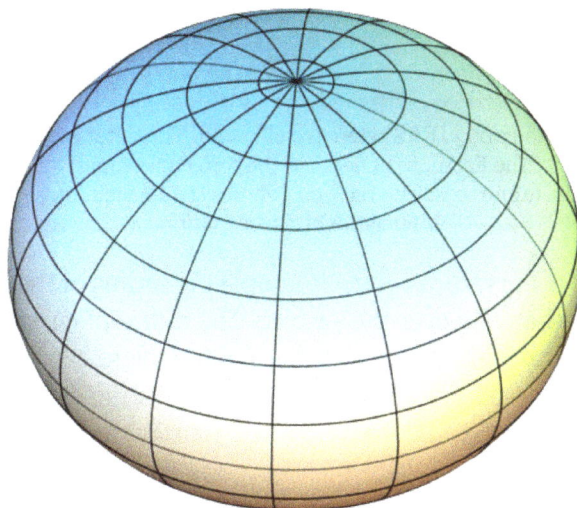

An oblate spheroid (highly exaggerated relative to the actual Earth)

Since the Earth is flattened at the poles and bulges at the equator, geodesy represents the shape of the earth with an oblate spheroid. The oblate spheroid, or oblate ellipsoid, is an ellipsoid of revolution obtained by rotating an ellipse about its shorter axis. It is the regular geometric shape that most nearly approximates the shape of the Earth. A spheroid describing the figure of the Earth or other celestial body is called a reference ellipsoid. The reference ellipsoid for Earth is called an Earth ellipsoid.

An ellipsoid of revolution is uniquely defined by two numbers: two dimensions, or one dimension and a number representing the difference between the two dimensions. Geodesists, by convention, use the semimajor axis and flattening. The size is represented by the radius at the equator (the semimajor axis of the cross-sectional ellipse) and designated by the letter a. The shape of the ellipsoid is given by the flattening, f, which indicates how much the ellipsoid departs from spherical. (In practice, the two defining numbers are usually the equatorial radius and the reciprocal of the flattening, rather than the flattening itself; for the WGS84 spheroid used by today's GPS systems, the reciprocal of the flattening is set at 298.257223563 exactly.)

The difference between a sphere and a reference ellipsoid for Earth is small, only about one part in 300. Historically, flattening was computed from grade measurements. Nowadays, geodetic networks and satellite geodesy are used. In practice, many reference ellipsoids have been developed over the centuries from different surveys. The flattening value varies slightly from one reference ellipsoid to another, reflecting local conditions and whether the reference ellipsoid is intended to model the entire Earth or only some portion of it.

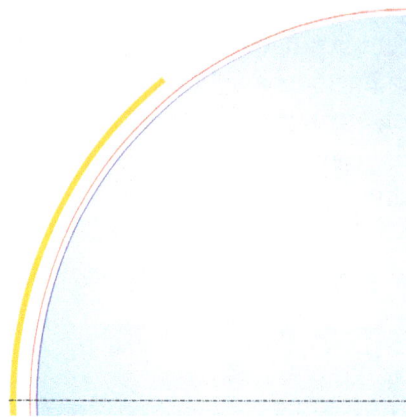

Scale drawing of the oblateness of the 2003 IERS reference ellipsoid. The outer edge of the dark blue line is an ellipse with the same eccentricity as that of the Earth, with north at the top. For comparison, the outer edge of light blue area is a circle of diameter equal to the minor axis. The red line denotes the Karman line and yellow area, the range of the International Space Station.

A sphere has a single radius of curvature, which is simply the radius of the sphere. More complex surfaces have radii of curvature that vary over the surface. The radius of curvature describes the radius of the sphere that best approximates the surface at that point. Oblate ellipsoids have constant radius of curvature east to west along parallels, if a graticule is drawn on the surface, but varying curvature in any other direction. For an oblate ellipsoid, the polar radius of curvature r_p is larger than the equatorial

$$r_p = \frac{a^2}{b},$$

because the pole is flattened: the flatter the surface, the larger the sphere must be to approximate it. Conversely, the ellipsoid's north-south radius of curvature at the equator r_e is smaller than the polar

$$r_e = \frac{b^2}{a}$$

where a is the distance from the center of the ellipsoid to the equator (semi-major axis), and b is the distance from the center to the pole. (semi-minor axis)

More Complicated Shapes

The possibility that the Earth's equator is an ellipse rather than a circle and therefore that the ellipsoid is triaxial has been a matter of scientific controversy for many years. Modern technological developments have furnished new and rapid methods for data collection and, since the launch of *Sputnik 1*, orbital data have been used to investigate the theory of ellipticity.

A second theory, more complicated than triaxiality, proposed that observed long periodic orbital variations of the first Earth satellites indicate an additional depression at the south pole accompanied by a bulge of the same degree at the north pole. It is also contended that the northern middle latitudes were slightly flattened and the southern middle latitudes bulged in a similar amount. This concept suggested a slightly pear-shaped Earth and was the subject of much public discussion. Modern geodesy tends to retain the ellipsoid of revolution and treat triaxiality and pear shape as a part of the geoid figure: they are represented by the spherical harmonic coefficients C_{22}, S_{22} and C_{30}, respectively, corresponding to degree and order numbers 2.2 for the triaxiality and 3.0 for the pear shape.

Geoid

It was stated earlier that measurements are made on the apparent or topographic surface of the Earth and it has just been explained that computations are performed on an ellipsoid. One other surface is involved in geodetic measurement: the geoid. In geodetic surveying, the computation of the geodetic coordinates of points is commonly performed on a reference ellipsoid closely approximating the size and shape of the Earth in the area of the survey. The actual measurements made on the surface of the Earth with certain instruments are however referred to the geoid. The ellipsoid is a mathematically defined regular surface with specific dimensions. The geoid, on the other hand, coincides with that surface to which the oceans would conform over the entire Earth if free to adjust to the combined effect of the Earth's mass attraction (gravitation) and the centrifugal force of the Earth's rotation. As a result of the uneven distribution of the Earth's mass, the geoidal surface is irregular and, since the ellipsoid is a regular surface, the separations between the two, referred to as geoid undulations, geoid heights, or geoid separations, will be irregular as well.

The geoid is a surface along which the gravity potential is everywhere equal and to which the direction of gravity is always perpendicular. The latter is particularly important because optical instruments containing gravity-reference leveling devices are commonly used to make geodetic measurements. When properly adjusted, the vertical axis of the instrument coincides with the direction of gravity and is, therefore, perpendicular to the geoid. The angle between the plumb line which is perpendicular to the geoid (sometimes called "the vertical") and the perpendicular to the ellipsoid (sometimes called "the ellipsoidal normal") is defined as the deflection of the vertical. It has two components: an east-west and a north-south component.

Earth Rotation and Earth's Interior

Determining the exact figure of the Earth is not only a geodetic operation or a task of geometry, but is also related to geophysics. Without any idea of the Earth's interior, we can state a "constant density" of 5.515 g/cm³ and, according to theoretical arguments. such a body rotating like the Earth would have a flattening of 1:230.

In fact, the measured flattening is 1:298.25, which is more similar to a sphere and a strong argument that the Earth's core is *very compact*. Therefore, the density must be a function of the depth, reaching from about 2.7 g/cm³ at the surface (rock density of granite, limestone etc.) up to approximately 15 within the inner core. Modern seismology yields a value of 16 g/cm³ at the center of the Earth.

Global and Regional Gravity Field

Also with implications for the physical exploration of the Earth's interior is the gravitational field, which can be measured very accurately at the surface and remotely by satellites. True vertical generally does not correspond to theoretical vertical (deflection ranges up to 50") because topography and all *geological masses* disturb the gravitational field. Therefore, the gross structure of the earth's crust and mantle can be determined by geodetic-geophysical models of the subsurface.

Volume

Earth's volume is approximately 1,083,210,000,000 km³ (2.5988×10¹¹ cu mi).

Geographical Distance

Geographical distance is the distance measured along the surface of the earth. The formulae in this chapter calculates distances between points which are defined by geographical coordinates in terms of latitude and longitude. This distance is an element in solving the second (inverse) geodetic problem.

Introduction

Calculating the distance between geographical coordinates is based on some level of abstraction; it does not provide an *exact* distance, which is unattainable if one attempted to account for every irregularity in the surface of the earth. Common abstractions for the surface between two geographic points are:

- Flat surface;
- Spherical surface;
- Ellipsoidal surface.

All abstractions above ignore changes in elevation. Calculation of distances which account for changes in elevation relative to the idealized surface are not discussed.

Nomenclature

Distance, D, is calculated between two points, P_1 and P_2. The geographical coordinates of the two points, as (latitude, longitude) pairs, are (ϕ_1, λ_1) and (ϕ_2, λ_2), respectively. Which of the two points is designated as P_1 is not important for the calculation of distance.

Latitude and longitude coordinates on maps are usually expressed in degrees. In the given forms of the formulae below, one or more values *must* be expressed in the specified units to obtain the correct result. Where geographic coordinates are used as the argument of a trigonometric function, the values may be expressed in any angular units compatible with the method used to determine the value of the trigonometric function. Many electronic calculators allow calculations of trigonometric functions in either degrees or radians. The calculator mode must be compatible with the units used for geometric coordinates.

Differences in latitude and longitude are labeled and calculated as follows:

$$\Delta\phi = \phi_2 - \phi_1;$$
$$\Delta\lambda = \lambda_2 - \lambda_1.$$

It is not important whether the result is positive or negative when used in the formulae below.

"Mean latitude" is labeled and calculated as follows:

$$\phi_m = \frac{\phi_1 + \phi_2}{2}.$$

Colatitude is labeled and calculated as follows:

For latitudes expressed in radians:

$$\theta = \frac{\pi}{2} - \phi;$$

For latitudes expressed in degrees:

$$\theta = 90^\circ - \phi.$$

Unless specified otherwise, the radius of the earth for the calculations below is:

R = 6,371.009 kilometers = 3,958.761 statute miles = 3,440.069 nautical miles.

D = Distance between the two points, as measured along the surface of the earth and in the same units as the value used for radius unless specified otherwise.

Singularities and Discontinuity of Latitude/Longitude

Longitude has singularities at the Poles (longitude is undefined) and a discontinuity at the ±180° meridian. Also, planar projections of the circles of constant latitude are highly curved near the Poles. Hence, the above equations for delta latitude/longitude ($\Delta\phi$, $\Delta\lambda$) and mean latitude (ϕ_m) may not give the expected answer for positions near the Poles or the ±180° meridian. Consider e.g. the value of $\Delta\lambda$ ("east displacement") when   and λ_2 are on either side of the ±180° meridian, or the value of ϕ_m ("mean latitude") for the two positions (ϕ_1 =89°, λ_1 =45°) and (ϕ_2 =89°, λ_2 =−135°).

If a calculation based on latitude/longitude should be valid for all Earth positions, it should be verified that the discontinuity and the Poles are handled correctly. Another solution is to use *n*-vector instead of latitude/longitude, since this representation does not have discontinuities or singularities.

Flat-surface Formulae

A planar approximation for the surface of the earth may be useful over small distances. The accuracy of distance calculations using this approximation become increasingly inaccurate as:

* The separation between the points becomes greater;

* A point becomes closer to a geographic pole.

The shortest distance between two points in plane is a straight line. The Pythagorean theorem is used to calculate the distance between points in a plane.

Even over short distances, the accuracy of geographic distance calculations which assume a flat Earth depend on the method by which the latitude and longitude coordinates have been projected onto the plane. The projection of latitude and longitude coordinates onto a plane is the realm of cartography.

The formulae presented provide varying degrees of accuracy.

Spherical Earth Projected to a Plane

This formula takes into account the variation in distance between meridians with latitude:

$$D = R\sqrt{(\Delta\phi)^2 + (\cos(\phi_m)\Delta\lambda)^2},$$

where:

$\Delta\phi$ and $\Delta\lambda$ are in radians;

ϕ_m must be in units compatible with the method used for determining $\cos(\phi_m)$.

To convert latitude or longitude to radians use

$$1^\circ = (\pi/180)\text{radians}.$$

This approximation is very fast and produces fairly accurate result for small distances. Also, when ordering locations by distance, such as in a database query, it is much faster to order by squared distance, eliminating the need for computing the square root.

Ellipsoidal Earth Projected to a Plane

The FCC prescribes essentially the following formulae in 47 CFR 73.208 for distances not exceeding 475 km /295 miles:

$$D = \sqrt{(K_1\Delta\phi)^2 + (K_2\Delta\lambda)^2},$$

where

D = Distance in kilometers;

$\Delta\phi$ and $\Delta\lambda$ are in degrees;

ϕ_m must be in units compatible with the method used for determining $\cos(\phi_m)$;

$$K_1 = 111.13209 - 0.56605\cos(2\phi_m) + 0.00120\cos(4\phi_m);$$
$$K_2 = 111.41513\cos(\phi_m) - 0.09455\cos(3\phi_m) + 0.00012\cos(5\phi_m).$$

It may be interesting to note that:

$$K_1 = M\frac{\pi}{180} = \text{kilometers per degree of latitude difference;}$$

$$K_2 = \cos(\phi_m)N\frac{\pi}{180} = \text{kilometers per degree of longitude difference;}$$

where M and N are the *meridional* and its perpendicular, or "*normal*", radii of curvature (the expressions in the FCC formula are derived from the binomial series expansion form of M and N, set to the *Clarke 1866* reference ellipsoid).

Polar Coordinate flat-Earth Formula

$$D = R\sqrt{\theta_1^2 + \theta_2^2 - 2\theta_1\theta_2\cos(\Delta\lambda)},$$

where the colatitude values are in radians. For a latitude measured in degrees, the colatitude in radians may be calculated as follows: $\theta = \frac{\pi}{180}(90° - \phi)$.

Spherical-surface Formulae

If we are willing to accept a possible error of 0.5%, we can use formulas of spherical trigonometry on the sphere that best approximates the surface of the earth.

The shortest distance along the surface of a sphere between two points on the surface is along the great-circle which contains the two points.

Tunnel Distance

A tunnel between points on Earth is defined by a line through three-dimensional space between the points of interest. The great circle chord length may be calculated as follows for the corresponding unit sphere:

$$\Delta X = \cos(\phi_2)\cos(\lambda_2) - \cos(\phi_1)\cos(\lambda_1);$$
$$\Delta Y = \cos(\phi_2)\sin(\lambda_2) - \cos(\phi_1)\sin(\lambda_1);$$
$$\Delta Z = \sin(\phi_2) - \sin(\phi_1);$$
$$C_h = \sqrt{(\Delta X)^2 + (\Delta Y)^2 + (\Delta Z)^2}.$$

The tunnel distance between points on the surface of a spherical Earth is $D = RC_h$. For short distances ($D \ll R$), this underestimates the great circle distance by $D(D/R)^2/24$.

Ellipsoidal-surface Formulae

An ellipsoid approximates the surface of the earth much better than a sphere or a flat surface does. The shortest distance along the surface of an ellipsoid between two points on the surface is along the geodesic. Geodesics follow more complicated paths than great circles and in particular, they usually don't return to their starting positions after one circuit of the earth. This is illustrated in the figure on the right where f is taken to be 1/50 to accentuate the effect. Finding the geodesic between two points on the earth, the so-called inverse geodetic problem, was the focus of many mathematicians and geodesists over the course of the 18th and 19th centuries with major contributions by Clairaut, Legendre, Bessel, and Helmert. Rapp provides a good summary of this work.

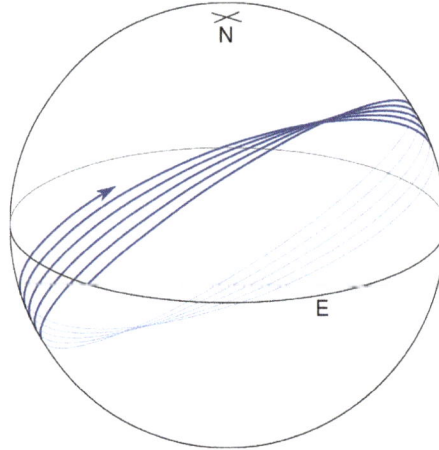

Geodesic on an oblate ellipsoid

Methods for computing the geodesic distance are widely available in geographical information systems, software libraries, standalone utilities, and online tools. The most widely used algorithm is by Vincenty, who uses a series which is accurate to third order in the flattening of the ellipsoid, i.e., about 0.5 mm; however, the algorithm fails to converge for points that are nearly anti-podal. This defect is cured in the algorithm given by Karney, who employs series which are accurate to sixth order in the flattening. This results in an algorithm which is accurate to full double precision and which converges for arbitrary pairs of points on the earth. This algorithm is implemented in GeographicLib.

The exact methods above are feasible when carrying out calculations on a computer. They are intended to give millimeter accuracy on lines of any length; we can use simpler formulas if we don't need millimeter accuracy, or if we do need millimeter accuracy but the line is short. Rapp, Chap. 6, describes the Puissant method, the Gauss mid-latitude method, and the Bowring method.

Lambert's Formula for Long Lines

Lambert's formulae give accuracy on the order of 10 meters over thousands of kilometers. First convert the latitudes ϕ_1, ϕ_2 of the two points to reduced latitudes β_1, β_2

$$\tan \beta = (1 - f) \tan \phi,$$

where f is the flattening. Then calculate the central angle σ in radians between two points (β_1, λ_1) and (β_2, λ_2) on a sphere in the usual way (law of cosines or haversine formula), with longitudes λ_1 and λ_2 being the same on the sphere as on the spheroid.

$$P = \frac{\beta_1 + \beta_2}{2} \qquad Q = \frac{\beta_2 - \beta_1}{2}$$

$$X = (\sigma - \sin \sigma) \frac{\sin^2 P \cos^2 Q}{\cos^2 \frac{\sigma}{2}} \qquad Y = (\sigma + \sin \sigma) \frac{\cos^2 P \sin^2 Q}{\sin^2 \frac{\sigma}{2}}$$

$$\text{distance} = a\left(\sigma - \tfrac{f}{2}(X + Y)\right)$$

where a is the equatorial radius of the chosen spheroid.

On the GRS 80 spheroid Lambert's formula is off by

0 North 0 West to 40 North 120 West, 12.6 meters

0N 0W to 40N 60W, 6.6 meters

40N 0W to 40N 60W, 0.85 meter

Bowring's Method for Short Lines

Bowring maps the points to a sphere of radius R', with latitude and longitude represented as φ' and λ'. Define

$$A = \sqrt{1 + e'^2 \cos^4 \phi_1}, \quad B = \sqrt{1 + e'^2 \cos^2 \phi_1},$$

where the second eccentricity squared is

$$e'^2 = \frac{a^2 - b^2}{b^2} = \frac{f(2 - f)}{(1 - f)^2}.$$

The spherical radius is

$$R' = \frac{\sqrt{1 + e'^2}}{B^2} a.$$

(The Gaussian curvature of the ellipsoid at φ_1 is $1/R'^2$.) The spherical coordinates are given by

$$\tan \phi'_1 = \frac{\tan \phi}{B},$$

$$\Delta \phi' = \frac{\Delta \phi}{B}\left[1 + \frac{3e'^2}{4B^2}(\Delta \phi)\sin(2\phi_1 + \tfrac{2}{3}\Delta \phi)\right],$$

$$\Delta \lambda' = A \Delta \lambda,$$

where $\Delta \phi = \phi_2 - \phi_1$, $\Delta \phi' = \phi'_2 - \phi'_1$, $\Delta \lambda = \lambda_2 - \lambda_1$, $\Delta \lambda' = \lambda'_2 - \lambda'_1$. The resulting problem on the sphere may be solved using the techniques for great-circle navigation to give approximations for the spheroidal distance and bearing.

Great-circle Distance

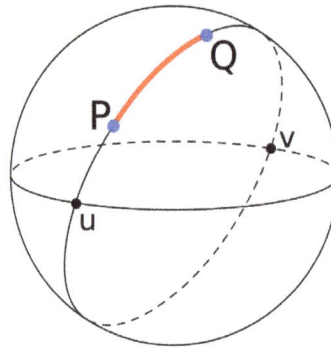

A diagram illustrating great-circle distance (drawn in red) between two
points on a sphere, P and Q. Two antipodal points, u and v, are also depicted.

The great-circle distance or orthodromic distance is the shortest distance between two points on
the surface of a sphere, measured along the surface of the sphere (as opposed to a straight line
through the sphere's interior). The distance between two points in Euclidean space is the length of
a straight line between them, but on the sphere there are no straight lines. In spaces with curvature, straight lines are replaced by geodesics. Geodesics on the sphere are the *great circles* (circles
on the sphere whose centers coincide with the center of the sphere).

Through any two points on a sphere that are not directly opposite each other, there is a unique
great circle. The two points separate the great circle into two arcs. The length of the shorter arc is
the great-circle distance between the points. A great circle endowed with such a distance is called
a Riemannian circle in Riemannian geometry.

Between two points that are directly opposite each other, called *antipodal points*, there are in-
finitely many great circles, but all great circle arcs between antipodal points have the same length,
i.e. half the circumference of the circle, or πr , where r is the radius of the sphere.

The Earth is nearly spherical. so great-circle distance formulas give the distance between points on
the surface of the Earth (*as the crow flies*) correct to within 0.5% or so.

Formulas

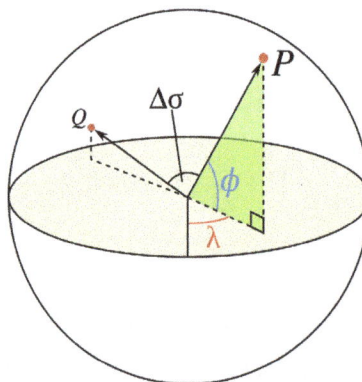

An illustration of the central angle, $\Delta\sigma$, between two points, P and Q. λ
and φ are the longitudinal and latitudinal angles of P respectively

Let ϕ_1, λ_1 and ϕ_2, λ_2 be the geographical latitude and longitude of two points 1 and 2, and $\Delta\phi, \Delta\lambda$ their absolute differences; then $\Delta\sigma$, the central angle between them, is given by the spherical law of cosines:

$$\Delta\sigma = \arccos\big(\sin\phi_1 \cdot \sin\phi_2 + \cos\phi_1 \cdot \cos\phi_2 \cdot \cos(\Delta\lambda)\big).$$

The distance d, i.e. the arc length, for a sphere of radius r and $\Delta\sigma$, given in radians

$$d = r\Delta\sigma.$$

Computational Formulas

On computer systems with low floating-point precision, the spherical law of cosines formula can have large rounding errors if the distance is small (if the two points are a kilometer apart on the surface of the Earth, the cosine of the central angle comes out 0.99999999). For modern 64-bit floating-point numbers, the spherical law of cosines formula, given above, does not have serious rounding errors for distances larger than a few meters on the surface of the Earth. The haversine formula is numerically better-conditioned for small distances:

$$\Delta\sigma = 2\arcsin\sqrt{\sin^2\left(\frac{\Delta\phi}{2}\right) + \cos\phi_1 \cdot \cos\phi_2 \cdot \sin^2\left(\frac{\Delta\lambda}{2}\right)}.$$

Historically, the use of this formula was simplified by the availability of tables for the haversine function: $\mathrm{hav}(\theta) = \sin^2(\theta/2)$.

Although this formula is accurate for most distances on a sphere, it too suffers from rounding errors for the special (and somewhat unusual) case of antipodal points (on opposite ends of the sphere). A more complicated formula that is accurate for all distances is the following special case of the Vincenty formula for an ellipsoid with equal major and minor axes:

$$\Delta\sigma = \arctan\frac{\sqrt{\big(\cos\phi_2 \cdot \sin(\Delta\lambda)\big)^2 + \big(\cos\phi_1 \cdot \sin\phi_2 - \sin\phi_1 \cdot \cos\phi_2 \cdot \cos(\Delta\lambda)\big)^2}}{\sin\phi_1 \cdot \sin\phi_2 + \cos\phi_1 \cdot \cos\phi_2 \cdot \cos(\Delta\lambda)}.$$

When programming a computer, one should use the atan2() function rather than the ordinary arctangent function (atan()), so that $\Delta\sigma$ is placed in the correct quadrant.

The determination of the great-circle distance is just part of the more general problem of great-circle navigation, which also computes the azimuths at the end points and intermediate way-points.

Vector Version

Another representation of similar formulas, but using normal vectors instead of latitude/longitude to describe the positions, is found by means of 3D vector algebra, i.e. utilizing the dot product, cross product, or a combination:

$$\Delta\sigma = \arccos(\mathbf{n}_1 \cdot \mathbf{n}_2)$$

$$\Delta\sigma = \arcsin|\mathbf{n}_1 \times \mathbf{n}_2|$$

$$\Delta\sigma = \arctan\frac{|\mathbf{n}_1 \times \mathbf{n}_2|}{\mathbf{n}_1 \cdot \mathbf{n}_2}$$

where $\mathbf{n}_1$ and $\mathbf{n}_2$ are the normals to the ellipsoid at the two positions 1 and 2. Similarly to the equations above based on latitude and longitude, the expression based on arctan is the only one that is well-conditioned for all angles. It should be noted that the expression based on arctan requires the magnitude of the cross product over the dot product.

From Chord Length

A line through three-dimensional space between points of interest on a spherical Earth is the chord of the great circle between the points. The central angle between the two points can be determined from the chord length. The great circle distance is proportional to the central angle.

The great circle chord length, C_h , may be calculated as follows for the corresponding unit sphere, by means of Cartesian subtraction:

$$\Delta X = \cos\phi_2 \cdot \cos\lambda_2 - \cos\phi_1 \cdot \cos\lambda_1;$$

$$\Delta Y = \cos\phi_2 \cdot \sin\lambda_2 - \cos\phi_1 \cdot \sin\lambda_1;$$

$$\Delta Z = \sin\phi_2 - \sin\phi_1;$$

$$C = \sqrt{(\Delta X)^2 + (\Delta Y)^2 + (\Delta Z)^2}$$

The central angle is:

$$\Delta\sigma = 2\arcsin\frac{C}{2}.$$

The great circle distance is:

$$d = r\Delta\sigma.$$

In this last formula, the central angle must be in radians. Alternatively, when working in nautical miles, the distance may be calculated directly by converting the central angle in degrees to minutes (i.e. multiply by 60).

Radius for Spherical Earth

The shape of the Earth closely resembles a flattened sphere (a spheroid) with equatorial radius a of 6378.137 km; distance b from the center of the spheroid to each pole is 6356.752 km. When calculating the length of a short north-south line at the equator, the circle that best approximates that line has a radius of b^2/a (which equals the meridian's semi-latus rectum), or 6335.439 km, while the spheroid at the poles is best approximated by a sphere of radius a^2/b , or 6399.594 km, a 1% difference. So long as we are assuming a spherical Earth, any single formula for distance on the Earth is only guaranteed correct within 0.5% (though we can do better if our formula is only intended to apply to a limited area). A good choice for the radius is the mean earth radius, $R_1 = \frac{1}{3}(2a + b) \approx 6371\text{km}$ (for the WGS84 ellipsoid); in the limit of small flattening, this choice minimizes the mean square relative error in the estimates for distance.

Metres Above Sea Level

Metres above mean sea level (MAMSL) or simply metres above sea level (MASL or m a.s.l.) is a standard metric measurement in metres of the elevation or altitude of a location in reference to a historic mean sea level. Mean sea levels are affected by climate change and other factors and change over time. For this and other reasons, recorded measurements of elevation above sea level might differ from the actual elevation of a given location over sea level at a given moment.

Uses

Metres above sea level is the standard measurement of the elevation or altitude of:

- Geographic locations such as towns, mountains and other landmarks.
- The top of buildings and other structures.
- Flying objects such as airplanes or helicopters.

How it is Determined

The elevation or altitude in metres above sea level of a location, object, or point can be determined in a number of ways. The most common include:

- Global Positioning System (GPS), which triangulates a location in reference to multiple satellites.
- Altimeters. They typically measure atmospheric pressure, which decreases as altitude increases.
- Aerial photography.
- Surveying.

Accurate measurement of historical mean sea levels is complex. Land mass subsidence (as occurs naturally in some islands) can give the appearance of rising sea levels. Conversely, markings on land masses that are uplifted due to geological processes can suggest a lowering of mean sea level.

Other Measurement Systems

Feet above sea level is the most common analogue for metres above sea level in the American measurement system.

Abbreviations

Metres above sea level is commonly abbreviated mamsl or MAMSL, based on the abbreviation AMSL for above mean sea level. Other abbreviations are m.a.s.l. and MASL.

Orthometric Height

The orthometric height of a point is the distance H along a plumb line from the point to the geoid.

Orthometric height is for practical purposes "height above sea level" but the current NAVD88 datum is tied to a defined elevation at one point rather than to any location's exact mean sea level.

Orthometric heights are usually used in the US for engineering work, although dynamic height may be chosen for large-scale hydrological purposes. Heights for measured points are shown on National Geodetic Survey data sheets, data that was gathered over many decades by precise spirit leveling over thousands of miles.

Since gravity is not constant over large areas the orthometric height of a level surface is not constant, and NGS orthometric heights are corrected for that effect. For example, gravity is 0.1% stronger in the northern United States than in the southern, so a level surface that has an orthometric height of 1000 meters in Montana will be 1001 meters high in Texas.

Practical applications must use a model rather than measurements to calculate the change in gravitational potential versus depth in the earth, since the geoid is below most of the land surface (e.g., the Helmert Orthometric heights of NAVD88).

GPS measurements give earth-centered coordinates, usually displayed as height above the reference ellipsoid, which cannot be related accurately to orthometric height above the geoid without accurate gravity data for that location. NGS is undertaking the GRAV-D ten-year program to obtain such data.

Alternatives to orthometric height include dynamic height and normal height.

Normal Height

Normal heights are heights above sea level, one of several types of height which are all computed slightly differently. Alternatives are: orthometric heights and dynamic heights.

The normal height $H*$ of a point is computed from geopotential numbers by dividing the point's geopotential number, i.e. its geopotential difference with that of sea level, by the average, normal gravity computed along the plumbline of the point. (More precisely, along the ellipsoidal normal, averaging over the height range from 0 — the ellipsoid — to $H*$, the procedure is thus recursive.

Normal heights are thus dependent upon the reference ellipsoid chosen. The Soviet Union and many other Eastern European countries have chosen a height system based on normal heights, determined by geodetic precise levelling. Normal gravity values are easy to compute and "hypothesis-free", i.e., one does not have to know, as one would for computing orthometric heights, the density of the Earth's crust around the plumbline.

Normal heights figure prominently in the theory of the Earth's gravity field developed by the school of M.S. Molodenskii. The reference surface that normal heights are measured from is called the quasi-geoid, a representation of "mean sea level" similar to the geoid and close to it, but lacking the physical interpretation of an equipotential surface.

Geopotential Height

Geopotential height is a vertical coordinate referenced to Earth's mean sea level — an adjustment to geometric height (elevation above mean sea level) using the variation of gravity with latitude and elevation. Thus it can be considered a "gravity-adjusted height". One usually speaks of the geopotential height of a certain pressure level, which would correspond to the geopotential height at which that pressure occurs.

Definition

At an elevation of h, the geopotential is defined as:

$$\Phi(h) = \int_0^h g(\phi, z)dz,$$

where $g(\phi, z)$ is the acceleration due to gravity, ϕ is latitude, and z is the geometric elevation. Thus geopotential is the gravitational potential energy per unit mass at that elevation h.

The geopotential height is:

$$Z_g(h) = \frac{\Phi(h)}{g_0},$$

which normalizes the geopotential to g_0, the standard gravity at mean sea level.

Usage

Geophysical scientists often use geopotential height as a function of pressure rather than pressure as a function of geometric height, because doing so in many cases makes analytical calculations more convenient. For example, the primitive equations which weather forecast models solve are more easily expressed in terms of geopotential than geometric height. Using the former eliminates air density from the equations.

A plot of geopotential height for a single pressure level shows the troughs and ridges, Highs and Lows, which are typically seen on upper air charts. The geopotential thickness between pressure levels — difference of the 850 hPa and 1000 hPa geopotential heights for example — is proportional to mean virtual temperature in that layer. Geopotential height contours can be used to calculate the geostrophic wind, which is faster where the contours are more closely spaced and tangential to the geopotential height contours.

The National Weather Service defines geopotential height as:

"...roughly the height above sea level of a pressure level. For example, if a station reports that the 500 mb [i.e. millibar] height at its location is 5600 m, it means that the level of the atmosphere over that station at which the atmospheric pressure is 500 mb is 5600 meters above sea level. This is an estimated height based on temperature and pressure data."

References

- "Meters above Sea Level - What does MSL stand for? Acronyms and abbreviations by The Free Online Dictionary.". Retrieved 2007-05-06

- Vanderwal, W; Wu, P; Sideris, M; Shum, C (2008). "Use of GRACE determined secular gravity rates for glacial isostatic adjustment studies in North-America". Journal of Geodynamics. 46 (3–5): 144. Bibcode:2008J-Geo...46..144V. doi:10.1016/j.jog.2008.03.007

- Fowler, C.M.R. (2005). The Solid Earth; An Introduction to Global Geophysics. United Kingdom: Cambridge University Press. p. 214. ISBN 9780521584098

- "WGS 84, N=M=180 Earth Gravitational Model". NGA: Office of Geomatics. National Geospatial-Intelligence Agency. Retrieved 17 December 2016

- Bessel, F. W. (1825). "Über die Berechnung der geographischen Langen und Breiten aus geodatischen Vermessungen". Astron. Nachr. 4 (86): 241–254. doi:10.1002/asna.201011352

- Paulson, Archie; Zhong, Shijie; Wahr, John (2007). "Inference of mantle viscosity from GRACE and relative sea level data". Geophysical Journal International. 171 (2): 497. Bibcode:2007GeoJI.171..497P. doi:10.1111/j.1365-246X.2007.03556.x

- Kopeikin, Sergei; Efroimsky, Michael; Kaplan, George (2009). Relativistic celestial mechanics of the solar system. Weinheim: Wiley-VCH. p. 704. ISBN 9783527408566

- "Earth's gravity definition". GRACE - Gravity Recovery and Climate Experiment. University of Texas at Austin. Retrieved 17 December 2016

- Karney, C. F. F. (2013). "Algorithms for geodesics". of Geodesy. 87 (1): 43–42. Bibcode:2013JGeod..87...43K. arXiv:1109.4448. doi:10.1007/s00190-012-0578-z (open access). Addenda

- Gade, Kenneth (2010). "A non-singular horizontal position representation" (PDF). The Journal of Navigation. Cambridge University Press. 63 (3): 395–417. doi:10.1017/S0373463309990415

- Wieczorek, M. A. (2007). "Gravity and Topography of the Terrestrial Planets". Treatise on Geophysics. p. 165. ISBN 9780444527486. doi:10.1016/B978-044452748-6.00156-5

- "Making maps compatible with GPS". Government of Ireland 1999. Archived from the original on 21 July 2011. Retrieved 15 April 2008

- Gade, Kenneth (2010). "A non-singular horizontal position representation" (PDF). The Journal of Navigation. Cambridge University Press. 63 (3): 395–417. doi:10.1017/S0373463309990415

- Evans, James (1998), The History and Practice of Ancient Astronomy, Oxford: Oxford University Press, pp. 102–103, ISBN 9780199874453

- Vincenty, T. (April 1975). "Direct and Inverse Solutions of Geodesics on the Ellipsoid with application of nested equations" (PDF). Survey Review. 23 (176): 88–93. doi:10.1179/sre.1975.23.176.88. Retrieved 2009-07-11. Addendum: Survey Review 23 (180): 294 (1976)

- McCaw, G. T. (1932). "Long lines on the Earth". Empire Survey Review. 1 (6): 259–263. doi:10.1179/sre.1932.1.6.259

4

Technologies used in Geodesy

The technologies related to geodesy are satellite navigation, global positioning system, Lidar and interferometric synthetic aperture radar. The system that uses satellites to locate places on Earth is known as satellite navigation. BeiDou navigation satellite system, Galileo and GLONASS are some of the examples of satellite navigation. Tools and techniques are an important component of any field of study. The following chapter elucidates the various tools and techniques that are related to geodesy.

Satellite Navigation

A satellite navigation or satnav system is a system that uses satellites to provide autonomous geo-spatial positioning. It allows small electronic receivers to determine their location (longitude, latitude, and altitude/elevation) to high precision (within a few metres) using time signals transmitted along a line of sight by radio from satellites. The system can be used for providing position, navigation or for tracking the position of something fitted with a receiver (satellite tracking). The signals also allow the electronic receiver to calculate the current local time to high precision, which allows time synchronisation. Satnav systems operate independently of any telephonic or internet reception, though these technologies can enhance the usefulness of the positioning information generated.

A satellite navigation system with global coverage may be termed a global navigation satellite system (GNSS). As of December 2016 only the United States NAVSTAR Global Positioning System (GPS), the Russian GLONASS and the European Union's Galileo are global operational GNSSs. The European Union's Galileo GNSS is scheduled to be fully operational by 2020. China is in the process of expanding its regional BeiDou Navigation Satellite System into the global BeiDou-2 GNSS by 2020. France, India and Japan are in the process of developing regional navigation and augmentation systems as well.

Global coverage for each system is generally achieved by a satellite constellation of 18–30 medium Earth orbit (MEO) satellites spread between several orbital planes. The actual systems vary, but use orbital inclinations of >50° and orbital periods of roughly twelve hours (at an altitude of about 20,000 kilometres or 12,000 miles).

Classification

Satellite navigation systems that provide enhanced accuracy and integrity monitoring usable for civil navigation are classified as follows:

- GNSS-1 is the first generation system and is the combination of existing satellite navigation systems (GPS and GLONASS), with Satellite Based Augmentation Systems (SBAS) or Ground Based Augmentation Systems (GBAS). In the United States, the satellite based component is the Wide Area Augmentation System (WAAS), in Europe it is the European Geostationary Navigation Overlay Service (EGNOS), and in Japan it is the Multi-Functional Satellite Augmentation System (MSAS). Ground based augmentation is provided by systems like the Local Area Augmentation System (LAAS).

- GNSS-2 is the second generation of systems that independently provides a full civilian satellite navigation system, exemplified by the European Galileo positioning system. These systems will provide the accuracy and integrity monitoring necessary for civil navigation; including aircraft. This system consists of L1 and L2 frequencies (in the L band of the radio spectrum) for civil use and L5 for system integrity. Development is also in progress to provide GPS with civil use L2 and L5 frequencies, making it a GNSS-2 system.

- Core Satellite navigation systems, currently GPS (United States), GLONASS (Russian Federation), Galileo (European Union) and Compass (China).

- Global Satellite Based Augmentation Systems (SBAS) such as Omnistar and StarFire.

- Regional SBAS including WAAS (US), EGNOS (EU), MSAS (Japan) and GAGAN (India).

- Regional Satellite Navigation Systems such as China's Beidou, India's NAVIC, and Japan's proposed QZSS.

- Continental scale Ground Based Augmentation Systems (GBAS) for example the Australian GRAS and the joint US Coast Guard, Canadian Coast Guard, US Army Corps of Engineers and US Department of Transportation National Differential GPS (DGPS) service.

- Regional scale GBAS such as CORS networks.

- Local GBAS typified by a single GPS reference station operating Real Time Kinematic (RTK) corrections.

History and Theory

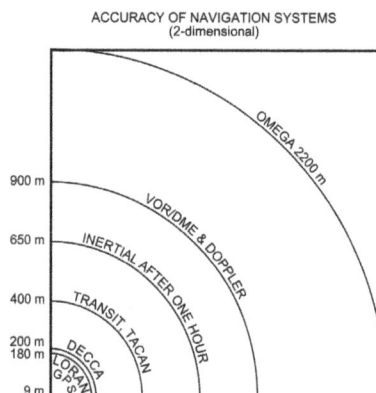

Ground based radio navigation has long been practiced. The DECCA, LORAN, GEE and Omega systems used terrestrial longwave radio transmitters which broadcast a radio pulse from a known "master" location, followed by a pulse repeated from a number of "slave" stations. The delay between the reception of the master signal and the slave signals allowed the receiver to deduce the distance to each of the slaves, providing a fix.

The first satellite navigation system was Transit, a system deployed by the US military in the 1960s. Transit's operation was based on the Doppler effect: the satellites travelled on well-known paths and broadcast their signals on a well-known radio frequency. The received frequency will differ slightly from the broadcast frequency because of the movement of the satellite with respect to the receiver. By monitoring this frequency shift over a short time interval, the receiver can determine its location to one side or the other of the satellite, and several such measurements combined with a precise knowledge of the satellite's orbit can fix a particular position.

Part of an orbiting satellite's broadcast included its precise orbital data. In order to ensure accuracy, the US Naval Observatory (USNO) continuously observed the precise orbits of these satellites. As a satellite's orbit deviated, the USNO would send the updated information to the satellite. Subsequent broadcasts from an updated satellite would contain its most recent ephemeris.

Modern systems are more direct. The satellite broadcasts a signal that contains orbital data (from which the position of the satellite can be calculated) and the precise time the signal was transmitted. The orbital ephemeris is transmitted in a data message that is superimposed on a code that serves as a timing reference. The satellite uses an atomic clock to maintain synchronization of all the satellites in the constellation. The receiver compares the time of broadcast encoded in the transmission of three (at sea level) or four different satellites, thereby measuring the time-of-flight to each satellite. Several such measurements can be made at the same time to different satellites, allowing a continual fix to be generated in real time using an adapted version of trilateration.

Each distance measurement, regardless of the system being used, places the receiver on a spherical shell at the measured distance from the broadcaster. By taking several such measurements and then looking for a point where they meet, a fix is generated. However, in the case of fast-moving receivers, the position of the signal moves as signals are received from several satellites. In addition, the radio signals slow slightly as they pass through the ionosphere, and this slowing varies with the receiver's angle to the satellite, because that changes the distance through the ionosphere. The basic computation thus attempts to find the shortest directed line tangent to four oblate spherical shells centred on four satellites. Satellite navigation receivers reduce errors by using combinations of signals from multiple satellites and multiple correlators, and then using techniques such as Kalman filtering to combine the noisy, partial, and constantly changing data into a single estimate for position, time, and velocity.

Civil and Military Uses

Satellite navigation using a laptop and a GPS receiver

The original motivation for satellite navigation was for military applications. Satellite navigation allows the precision in the delivery of weapons to targets, greatly increasing their lethality whilst reducing inadvertent casualties from mis-directed weapons. Satellite navigation also allows forces to be directed and to locate themselves more easily, reducing the fog of war.

The ability to supply satellite navigation signals is also the ability to deny their availability. The operator of a satellite navigation system potentially has the ability to degrade or eliminate satellite navigation services over any territory it desires.

Global Navigation Satellite Systems

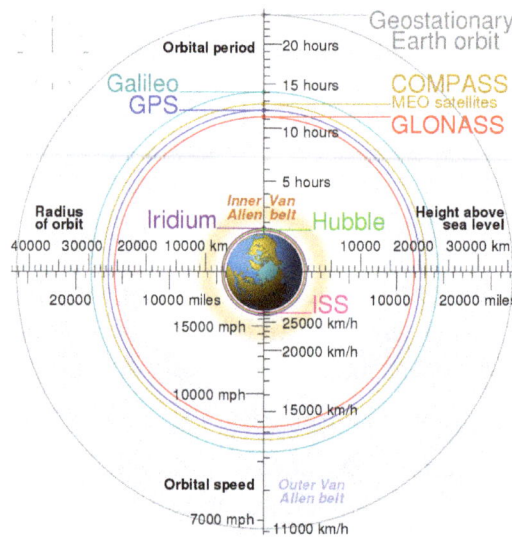

Comparison of geostationary, GPS, GLONASS, Galileo, Compass (MEO), International Space Station, Hubble Space Telescope and Iridium constellation orbits, with the Van Allen radiation belts and the Earth to scale. The Moon's orbit is around 9 times larger than geostationary orbit

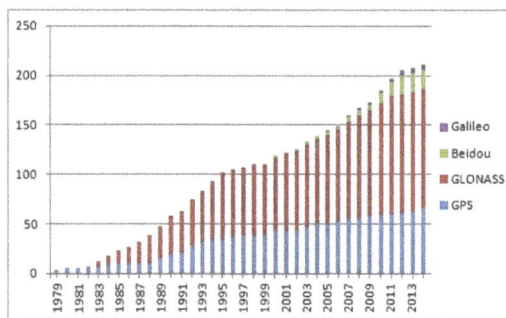

launched GNSS satellites 1978 to 2014

Operational

GPS

The United States' Global Positioning System (GPS) consists of up to 32 medium Earth orbit satellites in six different orbital planes, with the exact number of satellites varying as older satellites are retired and replaced. Operational since 1978 and globally available since 1994, GPS is currently the world's most utilized satellite navigation system.

GLONASS

The formerly Soviet, and now Russian, *Global'naya Navigatsionnaya Sputnikovaya Sistema* or GLONASS, is a space-based satellite navigation system that provides a civilian radionavigation-satellite service and is also used by the Russian Aerospace Defence Forces. The full orbital constellation of 24 GLONASS satellites enables full global coverage.

Galileo

The European Union and European Space Agency agreed in March 2002 to introduce their own alternative to GPS, called the Galileo positioning system. Galileo became operational on 15 December 2016 (global Early Operational Capability (EOC)) At an estimated cost of EUR 3.0 billion, the system of 30 MEO satellites was originally scheduled to be operational in 2010. The original year to become operational was 2014. The first experimental satellite was launched on 28 December 2005. Galileo is expected to be compatible with the modernized GPS system. The receivers will be able to combine the signals from both Galileo and GPS satellites to greatly increase the accuracy. Galileo is expected to be in full service in 2020 and at a substantially higher cost. The main modulation used in Galileo Open Service signal is the Composite Binary Offset Carrier (CBOC) modulation.

In Development

BeiDou-2

China has indicated they plan to complete the entire second generation Beidou Navigation Satellite System (BDS or BeiDou-2, formerly known as COMPASS), by expanding current regional (Asia-Pacific) service into global coverage by 2020. The BeiDou-2 system is proposed to consist of 30 MEO satellites and five geostationary satellites. A 16-satellite regional version (covering Asia and Pacific area) was completed by December 2012.

Regional Navigation Satellite Systems

BeiDou-1

Chinese regional (Asia-Pacific, 16 satellites) network to be expanded into the whole BeiDou-2 global system which consists of all 35 satellites by 2020.

NAVIC

The NAVIC or NAVigation with Indian Constellation is an autonomous regional satellite navigation system developed by Indian Space Research Organisation (ISRO) which would be under the total control of Indian government. The government approved the project in May 2006, with the intention of the system completed and implemented on 28 April 2016. It will consist of a constellation of 7 navigational satellites. 3 of the satellites will be placed in the Geostationary orbit (GEO) and the remaining 4 in the Geosynchronous orbit(GSO) to have a larger signal footprint and lower number of satellites to map the region. It is intended to provide an all-weather absolute position accuracy of better than 7.6 meters throughout India and within a region extending approximately

1,500 km around it. A goal of complete Indian control has been stated, with the space segment, ground segment and user receivers all being built in India. All seven satellites, IRNSS-1A, IRNSS-1B, IRNSS-1C, IRNSS-1D, IRNSS-1E, IRNSS-1F, and IRNSS-1G, of the proposed constellation were precisely launched on 1 July 2013, 4 April 2014, 16 October 2014, 28 March 2015, 20 January 2016, 10 March 2016 and 28 April 2016 respectively from Satish Dhawan Space Centre. The system is expected to be fully operational by August 2016.

QZSS

The Quasi-Zenith Satellite System (QZSS), is a proposed three-satellite regional time transfer system and enhancement for GPS covering Japan. The first demonstration satellite was launched in September 2010.

Augmentation

GNSS augmentation is a method of improving a navigation system's attributes, such as accuracy, reliability, and availability, through the integration of external information into the calculation process, for example, the Wide Area Augmentation System, the European Geostationary Navigation Overlay Service, the Multi-functional Satellite Augmentation System, Differential GPS, and inertial navigation systems.

DORIS

Doppler Orbitography and Radio-positioning Integrated by Satellite (DORIS) is a French precision navigation system. Unlike other GNSS systems, it is based on static emitting stations around the world, the receivers being on satellites, in order to precisely determine their orbital position. The system may be used also for mobile receivers on land with more limited usage and coverage. Used with traditional GNSS systems, it pushes the accuracy of positions to centimetric precision (and to millimetric precision for altimetric application and also allows monitoring very tiny seasonal changes of Earth rotation and deformations), in order to build a much more precise geodesic reference system.

Low Earth Orbit Satellite Phone Networks

The two current operational low Earth orbit satellite phone networks are able to track transceiver units with accuracy of a few kilometers using doppler shift calculations from the satellite. The coordinates are sent back to the transceiver unit where they can be read using AT commands or a graphical user interface. This can also be used by the gateway to enforce restrictions on geographically bound calling plans.

Examples of Satellite Navigation Systems

BeiDou Navigation Satellite System

The BeiDou Navigation Satellite System) is a Chinese satellite navigation system. It consists of two separate satellite constellations – a limited test system that has been operating since 2000, and a full-scale global navigation system that is currently under construction.

The first BeiDou system, officially called the BeiDou Satellite Navigation Experimental System and also known as BeiDou-1, consists of three satellites and offers limited coverage and applications. It has been offering navigation services, mainly for customers in China and neighboring regions, since 2000.

The second generation of the system, officially called the BeiDou Navigation Satellite System (BDS) and also known as COMPASS or BeiDou-2, will be a global satellite navigation system consisting of 35 satellites, and is under construction as of January 2015. It became operational in China in December 2011, with 10 satellites in use, and began offering services to customers in the Asia-Pacific region in December 2012. It is planned to begin serving global customers upon its completion in 2020.

In-mid 2015, China started the build-up of the third generation BeiDou system (BDS-3) in the global coverage constellation. The first BDS-3 satellite was launched 30 September 2015. As of March 2016, 4 BDS-3 in-orbit validation satellites have been launched.

According to China Daily, fifteen years after the satellite system was launched, it is now generating a turnover of $31.5 billion per annum for major companies such as China Aerospace Science and Industry Corp, AutoNavi Holdings Ltd, and China North Industries Group Corp.

Beidou has been described as a potential navigation satellite system to overtake GPS in global usage, and is expected to be more accurate than the GPS once it is fully completed.

Nomenclature

The official English name of the system is *BeiDou Navigation Satellite System*. It is named after the Big Dipper constellation, which is known in Chinese as *Běidǒu*. The name literally means "Northern Dipper", the name given by ancient Chinese astronomers to the seven brightest stars of the Ursa Major constellation. Historically, this set of stars was used in navigation to locate the North Star Polaris. As such, the name BeiDou also serves as a metaphor for the purpose of the satellite navigation system.

History

Conception and Initial Development

The original idea of a Chinese satellite navigation system was conceived by Chen Fangyun and his colleagues in the 1980s. According to the China National Space Administration, the development of the system would be carried out in three steps:

1. 2000–2003: experimental BeiDou navigation system consisting of 3 satellites

2. by 2012: regional BeiDou navigation system covering China and neighboring regions

3. by 2020: global BeiDou navigation system

The first satellite, *BeiDou-1A*, was launched on 30 October 2000, followed by *BeiDou-1B* on 20 December 2000. The third satellite, *BeiDou-1C* (a backup satellite), was put into orbit on 25 May 2003. The successful launch of *BeiDou-1C* also meant the establishment of the BeiDou-1 navigation system.

On 2 November 2006, China announced that from 2008 BeiDou would offer an open service with an accuracy of 10 meters, timing of 0.2 microseconds, and speed of 0.2 meters/second.

In February 2007, the fourth and last satellite of the BeiDou-1 system, *BeiDou-1D* (sometimes called *BeiDou-2A*, serving as a backup satellite), was sent up into space. It was reported that the satellite had suffered from a control system malfunction but was then fully restored.

In April 2007, the first satellite of BeiDou-2, namely *Compass-M1* (to validate frequencies for the BeiDou-2 constellation) was successfully put into its working orbit. The second BeiDou-2 constellation satellite *Compass-G2* was launched on 15 April 2009. On 15 January 2010, the official website of the BeiDou Navigation Satellite System went online, and the system's third satellite (*Compass-G1*) was carried into its orbit by a Long March 3C rocket on 17 January 2010. On 2 June 2010, the fourth satellite was launched successfully into orbit. The fifth orbiter was launched into space from Xichang Satellite Launch Center by an LM-3I carrier rocket on 1 August 2010. Three months later, on 1 November 2010, the sixth satellite was sent into orbit by LM-3C. Another satellite, the Beidou-2/Compass IGSO-5 (fifth inclined geosynchonous orbit) satellite, was launched from the Xichang Satellite Launch Center by a Long March-3A on 1 December 2011 (UTC).

Chinese Involvement in Galileo System

In September 2003, China intended to join the European Galileo positioning system project and was to invest €230 million (USD296 million, GBP160 million) in Galileo over the next few years. At the time, it was believed that China's "BeiDou" navigation system would then only be used by its armed forces. In October 2004, China officially joined the Galileo project by signing the *Agreement on the Cooperation in the Galileo Program between the "Galileo Joint Undertaking" (GJU) and the "National Remote Sensing Centre of China" (NRSCC)*. Based on the Sino-European Cooperation Agreement on Galileo program, China Galileo Industries (CGI), the prime contractor of the China's involvement in Galileo programs, was founded in December 2004. By April 2006, eleven cooperation projects within the Galileo framework had been signed between China and EU. However, the Hong Kong-based *South China Morning Post* reported in January 2008 that China was unsatisfied with its role in the Galileo project and was to compete with Galileo in the Asian market.

Phase III

- In November 2014, Beidou became part of the World-Wide Radionavigation System (WWRNS) at the 94th meeting of The International Maritime Organization (IMO) Maritime Safety Committee, which approved the "Navigation Safety Circular" of the Beidou Navigation Satellite System (BDS).

- At Beijing time 21:52, March 30, 2015, the first new-generation BeiDou Navigation satellite (and the 17th overall) was successfully set to orbit by a Long March 3C rocket.

Experimental System (BeiDou-1)

Description

BeiDou-1 is an experimental regional navigation system, which consists of four satellites (three working satellites and one backup satellite). The satellites themselves were based on the Chinese DFH-3 geostationary communications satellite and had a launch weight of 1,000 kilograms (2,200 pounds) each.

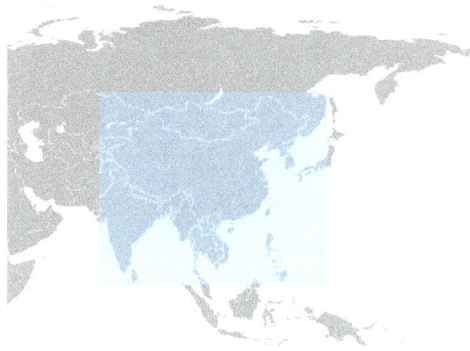

Coverage polygon of BeiDou-1.

Unlike the American GPS, Russian GLONASS, and European Galileo systems, which use medium Earth orbit satellites, BeiDou-1 uses satellites in geostationary orbit. This means that the system does not require a large constellation of satellites, but it also limits the coverage to areas on Earth where the satellites are visible. The area that can be serviced is from longitude 70°E to 140°E and from latitude 5°N to 55°N. A frequency of the system is 2491.75 MHz.

Completion

The first satellite, BeiDou-1A, was launched on October 31, 2000. The second satellite, BeiDou-1B, was successfully launched on December 21, 2000. The last operational satellite of the constellation, BeiDou-1C, was launched on May 25, 2003.

Position Calculation

In 2007, the official Xinhua News Agency reported that the resolution of the BeiDou system was as high as 0.5 metres. With the existing user terminals it appears that the calibrated accuracy is 20m (100m, uncalibrated).

Terminals

In 2008, a BeiDou-1 ground terminal cost around CN¥ 20,000RMB (US$2,929), almost 10 times the price of a contemporary GPS terminal. The price of the terminals was explained as being due to the cost of imported microchips. At the China High-Tech Fair ELEXCON of November 2009 in Shenzhen, a BeiDou terminal priced at CN¥ 3,000RMB was presented.

Applications

- Over 1,000 BeiDou-1 terminals were used after the 2008 Sichuan earthquake, providing information from the disaster area.

- As of October 2009, all Chinese border guards in Yunnan are equipped with BeiDou-1 devices.

According to Sun Jiadong, the chief designer of the navigation system, "Many organizations have been using our system for a while, and they like it very much."

The new-generation BeiDou satellites support short message service.

Global System (BeiDou Navigation Satellite System or BeiDou-2)

Description

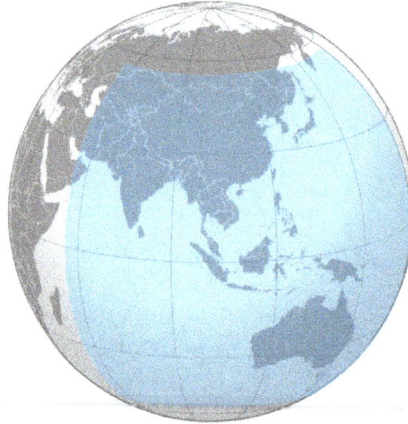

Coverage polygon of BeiDou-2 in 2012.

BeiDou-2 (formerly known as COMPASS) is not an extension to the older BeiDou-1, but rather supersedes it outright. The new system will be a constellation of 35 satellites, which include 5 geostationary orbit satellites for backward compatibility with BeiDou-1, and 30 non-geostationary satellites (27 in medium Earth orbit and 3 in inclined geosynchronous orbit), that will offer complete coverage of the globe.

The ranging signals are based on the CDMA principle and have complex structure typical of Galileo or modernized GPS. Similar to the other global navigation satellite systems (GNSSs), there will be two levels of positioning service: open (public) and restricted (military). The public service will be available globally to general users. When all the currently planned GNSSs are deployed, users of multi-constellation receivers will benefit from a total over 100 satellites, which will significantly improve all aspects of positioning, especially availability of the signals in so-called urban canyons. The general designer of the COMPASS navigation system is Sun Jiadong, who is also the general designer of its predecessor, the original BeiDou navigation system.

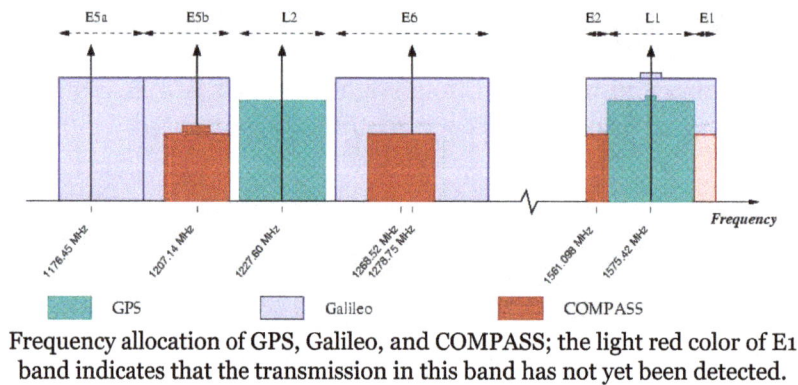

Frequency allocation of GPS, Galileo, and COMPASS; the light red color of E1 band indicates that the transmission in this band has not yet been detected.

Accuracy

There are two levels of service provided — a free service to civilians and licensed service to the Chinese government and military. The free civilian service has a 10-meter location-tracking accuracy, synchronizes clocks with an accuracy of 10 nanoseconds, and measures speeds to within 0.2 m/s. The restricted military service has a location accuracy of 10 centimetres, can be used for communication, and will supply information about the system status to the user. To date, the military service has been granted only to the People's Liberation Army and to the Military of Pakistan.

Frequencies

Frequencies for COMPASS are allocated in four bands: E1, E2, E5B, and E6 and overlap with Galileo. The fact of overlapping could be convenient from the point of view of the receiver design, but on the other hand raises the issues of inter-system interference, especially within E1 and E2 bands, which are allocated for Galileo's publicly regulated service. However, under International Telecommunication Union (ITU) policies, the first nation to start broadcasting in a specific frequency will have priority to that frequency, and any subsequent users will be required to obtain permission prior to using that frequency, and otherwise ensure that their broadcasts do not interfere with the original nation's broadcasts. It now appears that Chinese COMPASS satellites will start transmitting in the E1, E2, E5B, and E6 bands before Europe's Galileo satellites and thus have primary rights to these frequency ranges.

Although little was officially announced by Chinese authorities about the signals of the new system, the launch of the first COMPASS satellite permitted independent researchers not only to study general characteristics of the signals, but even to build a COMPASS receiver.

Compass-M1

Compass-M1 is an experimental satellite launched for signal testing and validation and for the frequency filing on 14 April 2007. The role of Compass-M1 for Compass is similar to the role of the GIOVE satellites for the Galileo system. The orbit of Compass-M1 is nearly circular, has an altitude of 21,150 km and an inclination of 55.5 degrees.

Compass-M1 transmits in 3 bands: E2, E5B, and E6. In each frequency band two coherent sub-signals have been detected with a phase shift of 90 degrees (in quadrature). These signal components are further referred to as "I" and "Q". The "I" components have shorter codes and are likely to be intended for the open service. The "Q" components have much longer codes, are more interference resistive, and are probably intended for the restricted service.

The investigation of the transmitted signals started immediately after the launch of Compass -M1 on 14 April 2007. Soon after in June 2007, engineers at CNES reported the spectrum and structure of the signals. A month later, researchers from Stanford University reported the complete decoding of the "I" signals components. The knowledge of the codes allowed a group of engineers at Septentrio to build the COMPASS receiver and report tracking and multipath characteristics of the "I" signals on E2 and E5B.

Characteristics of Compass Signals Reported as of May 2008 Compared to GPS-L1CA							
Parameters	**E2-I**	**E2-Q**	**E5B-I**	**E5B-Q**	**E6-I**	**E6-Q**	**GPS L1-CA**
Native notation	B1	B1	B2	B2	B3	B3	---

Code modulation	BPSK(2)	BPSK(2)	BPSK(2)	BPSK(10)	BPSK(10)	BPSK (10)	BPSK (1)
Carrier frequency, MHz	1561.098	1561.098	1207.14	1207.14	1268.52	1268.52	1575.42
Chip rate, Mchips/s	2.046	2.046	2.046	10.230	10.230	10.230	1.023
Code period, chips	2046	??	2046	??	10230	??	1023
Code period, ms	1.0	>400	1.0	>160	1.0	>160	1.0
Symbols/s	50	??	50	??	50	??	50
Navigation frames, s	6	??	6	??	??	??	6
Navigation sub-frames, s	30	??	30	??	??	??	30
Navigation period, min	12.0	??	12.0	??	??	??	12.5

Characteristics of the "I" signals on E2 and E5B are generally similar to the civilian codes of GPS (L1-CA and L2C), but Compass signals have somewhat greater power. The notation of Compass signals used in this page follows the naming of the frequency bands and agrees with the notation used in the American literature on the subject, but the notation used by the Chinese seems to be different and is quoted in the first row of the table.

Operation

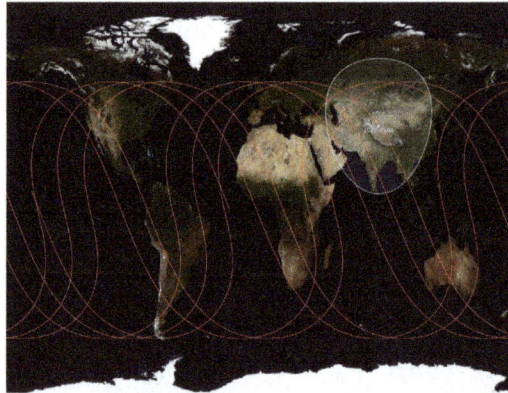

Ground track of BeiDou-M5 (2012-050A)

In December 2011, the system went into operation on a trial basis. It has started providing navigation, positioning and timing data to China and the neighbouring area for free from 27 December. During this trial run, Compass will offer positioning accuracy to within 25 meters, but the precision will improve as more satellites are launched. Upon the system's official launch, it pledged to offer general users positioning information accurate to the nearest 10 m, measure speeds within 0.2 m per second, and provide signals for clock synchronisation accurate to 0.02 microseconds.

The BeiDou-2 system began offering services for the Asia-Pacific region in December 2012. At this time, the system could provide positioning data between longitude 55°E to 180°E and from latitude 55°S to 55°N.

Completion

In December 2011, Xinhua stated that "the basic structure of the Beidou system has now been established, and engineers are now conducting comprehensive system test and evaluation. The system will provide test-run services of positioning, navigation and time for China and the neighboring areas before the end of this year, according to the authorities." The system became operational in the China region that same month. The global navigation system should be finished by 2020. As of December 2012, 16 satellites for BeiDou-2 have been launched, 14 of them are in service.

Constellation

Summary of Satellites					
Block	Launch Period	Satellite launches			Currently in orbit and healthy
		Success	Failure	Planned	
1	2000–2007	4	0	0	0
2	2007-2012	16	0	0	13
3	From 2015	7	0	18	7
Total		27	0	18	20

The regional Beidou-1 system was decommissioned at the end of 2012.

The first satellite of the second-generation system, Compass-M1 was launched in 2007. It was followed by further nine satellites during 2009-2011, achieving functional regional coverage. A total of 16 satellites were launched during this phase.

In 2015, the system began its transition towards global coverage with the first launch of a new-generation of satellites, and the 17th one within the new system. On July 25, 2015, the 18th and 19th satellites were successfully launched from the Xichang Satellite Launch Center, marking the first time for China to launch two satellites at once on top of a Long March 3B/Expedition-1 carrier rocket. The Expedition-1 is an independent upper stage capable of delivering one or more spacecraft into different orbits. On November 29, the 20th satellite was launched, carrying a hydrogen maser for the first time within the system.

In 2016, the 21st, 22nd and 23rd satellites were launched from Xichang Satellite Launch Center, the last two of which entered into service on August 5 and November 30, respectively.

GLONASS

GLONASS or "Global Navigation Satellite System", is a space-based satellite navigation system operating in the radionavigation-satellite service. It provides an alternative to GPS and is the second navigational system in operation with global coverage and of comparable precision.

Manufacturers of GPS devices say that adding GLONASS made more satellites available to them, meaning positions can be fixed more quickly and accurately, especially in built-up areas where the view to some GPS satellites is obscured by buildings. Smartphones generally tend to use the same chipsets and the versions used since 2015 receive GLONASS signals and positioning information along with GPS. Since 2012, GLONASS was the second most used positioning system in mobile phones after GPS. The system has the advantage that smartphone users receive a more accurate reception identifying location to within 2 meters.

A model of a GLONASS-K satellite displayed at CeBit 2011

Development of GLONASS began in the Soviet Union in 1976. Beginning on 12 October 1982, numerous rocket launches added satellites to the system until the constellation was completed in 1995. After a decline in capacity during the late 1990s, in 2001, under Vladimir Putin's presidency, the restoration of the system was made a top government priority and funding was substantially increased. GLONASS is the most expensive program of the Russian Federal Space Agency, consuming a third of its budget in 2010.

By 2010, GLONASS had achieved 100% coverage of Russia's territory and in October 2011, the full orbital constellation of 24 satellites was restored, enabling full global coverage. The GLONASS satellites' designs have undergone several upgrades, with the latest version being GLONASS-K.

System Description

GLONASS is a global satellite navigation system, providing real time position and velocity determination for civilian users. The satellites are located in middle circular orbit at 19,100 kilometres (11,900 mi) altitude with a 64.8 degree inclination and a period of 11 hours and 15 minutes. GLONASS' orbit makes it especially suited for usage in high latitudes (north or south), where getting a GPS signal can be problematic. The constellation operates in three orbital planes, with eight evenly spaced satellites on each. A fully operational constellation with global coverage consists of 24 satellites, while 18 satellites are necessary for covering the territory of Russia. To get a position fix the receiver must be in the range of at least four satellites.

Signal

FDMA

GLONASS satellites transmit two types of signal: open standard-precision signal L1OF/L2OF, and obfuscated high-precision signal L1SF/L2SF.

One of first samples Russian military rugged, combined GLONASS/GPS receiver, 2003 yr

The signals use similar DSSS encoding and binary phase-shift keying (BPSK) modulation as in GPS signals. All GLONASS satellites transmit the same code as their standard-precision signal; however each transmits on a different frequency using a 15-channel frequency division multiple access (FDMA) technique spanning either side from 1602.0 MHz, known as the L1 band. The center frequency is 1602 MHz + n × 0.5625 MHz, where n is a satellite's frequency channel number ($n=-7,-6,-5,...0,...,6$, previously $n=0,...,13$). Signals are transmitted in a 38° cone, using right-hand circular polarization, at an EIRP between 25 and 27 dBW (316 to 500 watts). Note that the 24-satellite constellation is accommodated with only 15 channels by using identical frequency channels to support antipodal (opposite side of planet in orbit) satellite pairs, as these satellites are never both in view of an earth-based user at the same time.

The L2 band signals use the same FDMA as the L1 band signals, but transmit straddling 1246 MHz with the center frequency 1246 MHz + n×0.4375 MHz, where n spans the same range as for L1. In the original GLONASS design, only obfuscated high-precision signal was broadcast in the L2 band, but starting with GLONASS-M, an additional civil reference signal L2OF is broadcast with an identical standard-precision code to the L1OF signal.

The open standard-precision signal is generated with modulo-2 addition (XOR) of 511 kbit/s pseudo-random ranging code, 50 bit/s navigation message, and an auxiliary 100 Hz meander sequence (Manchester code), all generated using a single time/frequency oscillator. The pseudo-random code is generated with a 9-stage shift register operating with a period of 1 ms.

The navigational message is modulated at 50 bits per second. The superframe of the open signal is 7500 bits long and consists of 5 frames of 30 seconds, taking 150 seconds (2.5 minutes) to transmit the continuous message. Each frame is 1500 bits long and consists of 15 strings of 100 bits (2 seconds for each string), with 85 bits (1.7 seconds) for data and check-sum bits, and 15 bits (0.3 seconds) for time mark. Strings 1-4 provide immediate data for the transmitting satellite, and are repeated every frame; the data include ephemeris, clock and frequency offsets, and satellite status. Strings 5-15 provide non-immediate data (i.e. almanac) for each satellite in the constellation, with frames I-IV each describing five satellites, and frame V describing remaining four satellites. The ephemerides are updated every 30 minutes using data from the Ground Control segment; they use Earth Centred Earth Fixed (ECEF) Cartesian coordinates in position and velocity, and include lunisolar acceleration parameters. The almanac uses modified Keplerian parameters and is updated daily.

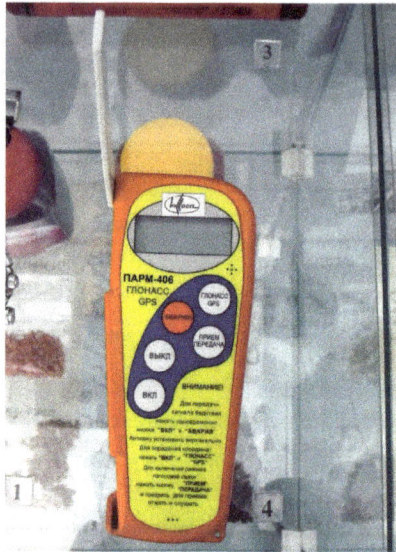

A combined GLONASS/GPS Personal Radio Beacon

The more accurate high-precision signal is available for authorized users, yet unlike the US P(Y) code, which is modulated by an encrypting W code, the GLONASS restricted-use codes are broadcast in the clear using only *security through obscurity*. The details of the high-precision signal have not been disclosed. The modulation (and therefore the tracking strategy) of the data bits on the L2SF code has recently changed from unmodulated to 250 bit/s burst at random intervals. The L1SF code is modulated by the navigation data at 50 bit/s without a Manchester meander code.

The high-precision signal is broadcast in phase quadrature with the standard-precision signal, effectively sharing the same carrier wave, but with a ten-times-higher bandwidth than the open signal. The message format of the high-precision signal remains unpublished, although attempts at reverse-engineering indicate that the superframe is composed of 72 frames, each containing 5 strings of 100 bits and taking 10 seconds to transmit, with total length of 36 000 bits or 720 seconds (12 minutes) for the whole navigational message. The additional data are seemingly allocated to critical Luni-Solar acceleration parameters and clock correction terms.

Accuracy

At peak efficiency, the standard-precision signal offers horizontal positioning accuracy within 5–10 meters, vertical positioning within 15 metres (49 ft), a velocity vector measuring within 10 centimetres per second (3.9 in/s), and timing within 200 ns, all based on measurements from four first-generation satellites simultaneously; newer satellites such as GLONASS-M improve on this.

GLONASS uses a coordinate datum named "PZ-90" (Earth Parameters 1990 – Parametry Zemli 1990), in which the precise location of the North Pole is given as an average of its position from 1990 to 1995. This is in contrast to the GPS's coordinate datum, WGS 84, which uses the location of the North Pole in 1984. As of 17 September 2007 the PZ-90 datum has been updated to version PZ-90.02 which differ from WGS 84 by less than 40 cm (16 in) in any given direction. Since 31 December 2013, version PZ-90.11 is being broadcast, which is aligned to the International Terrestrial Reference System at epoch 2011.0 at the centimeter level.

CDMA

Since 2008, new CDMA signals are being researched for use with GLONASS.

The interface control documents for GLONASS CDMA signals was published in August 2016.

According to GLONASS developers, there will be three open and two restricted CDMA signals. The open signal L3OC is centered at 1202.025 MHz and uses BPSK(10) modulation for both data and pilot channels; the ranging code transmits at 10.23 million chips per second, modulated onto the carrier frequency using QPSK with in-phase data and quadrature pilot. The data is error-coded with 5-bit Barker code and the pilot with 10-bit Neuman-Hoffman code.

Open L1OC and restricted L1SC signals are centered at 1600.995 MHz, and open L2OC and restricted L2SC signals are centered at 1248.06 MHz, overlapping with GLONASS FDMA signals. Open signals L1OC and L2OC use time-division multiplexing to transmit pilot and data signals, with BPSK(1) modulation for data and BOC(1,1) modulation for pilot; wide-band restricted signals L1SC and L2SC use BOC (5, 2.5) modulation for both data and pilot, transmitted in quadrature phase to the open signals; this places peak signal strength away from the center frequency of narrow-band open signals.

Binary phase-shift keying (BPSK) is used by standard GPS and GLONASS signals, however both BPSK and quadrature phase-shift keying (QPSK) can be considered as variations of quadrature amplitude modulation (QAM), specifically QAM-2 and QAM-4. Binary offset carrier (BOC) is the modulation used by Galileo, modernized GPS, and COMPASS.

The navigational message of CMDA signals is transmitted as a sequence of text strings. The message has variable size - each pseudo-frame usually includes six strings and contains ephemerides for the current satellite (string types 10, 11, and 12 in a sequence) and part of the almanac for three satellites (three strings of type 20). To transmit the full almanac for all current 24 satellites, a superframe of 8 pseudo-frames is required. In the future, the superframe will be expanded to 10 pseudo-frames of data to cover full 30 satellites. The message can also contain Earth rotation parameters, ionosphere models, long-term orbit parameters for GLONASS satellites, and COSPAS-SARSAT messages. The system time marker is transmitted with each string; UTC leap second correction is achieved by shortening or lengthening (zero-padding) the final string of the day by one second, with abnormal strings being discarded by the receiver. The strings have a version tag to facilitate forward compatibility: future upgrades to the message format will not break older equipment, which will continue to work by ignoring new data (as long as the constellation still transmits old string types), but up-to-date equipment will be able to use additional information from newer satellites.

The navigational message of the L3OC signal is transmitted at 100 bit/s, with each string of symbols taking 3 seconds (300 bits). A pseudo-frame of 6 strings takes 18 seconds (1800 bits) to transmit. A superframe of 8 pseudo-frames is 14400 bits long and takes 144 seconds (2 minutes 24 seconds) to transmit the full almanac.

The navigational message of the L1OC signal is transmitted at 100 bit/s. The string is 250 bits long and takes 2.5 seconds to transmit. A pseudo-frame is 1500 bits (15 seconds) long, and a superframe is 12000 bits or 120 seconds (2 minutes).

Glonass-K1 test satellite launched in 2011 introduced L3OC signal. Glonass-M satellites produced since 2014 (s/n 755+) will also transmit L3OC signal for testing purposes.

Enhanced Glonass-K1 and Glonass-K2 satellites, to be launched from 2018, will feature a full suite of modernized CDMA signals in the existing L1 and L2 bands, which includes L1SC, L1OC, L2SC, and L2OC, as well as the L3OC signal. Glonass-K series should gradually replace existing satellites starting from 2018, when Glonass-M launches will cease.

Glonass-KM satellites will be launched by 2025. Additional open signals are being studied for these satellites, based on the same frequencies and formats as GPS signals L5 and L1C and corresponding Galileo/COMPASS signals E1, E5a and E5b. These signals include:

- The open signal L1OCM will use BOC(1,1) modulation centered at 1575.42 MHz, similar to modernized GPS signal L1C and Galileo/COMPASS signal E1;

- The open signal L5OCM will use BPSK(10) modulation centered at 1176.45 MHz, similar to the GPS "Safety of Life" (L5) and Galileo/COMPASS signal E5a;

- The open signal L3OCM will use BPSK(10) modulation centered at 1207.14 MHz, similar to Galileo/COMPASS signal E5b.

Such an arrangement will allow easier and cheaper implementation of multi-standard GNSS receivers.

With the introduction of CDMA signals, the constellation will be expanded to 30 active satellites by 2025; this may require eventual deprecation of FDMA signals. The new satellites will be deployed into three additional planes, bringing the total to six planes from the current three—aided by System for Differential Correction and Monitoring (SDCM), which is a GNSS augmentation system based on a network of ground-based control stations and communication satellites Luch 5A and Luch 5B. Additional satellites may use Molniya orbit, Tundra orbit, geosynchronous orbit, or inclined orbit to offer increased regional availability, similar to Japanese QZSS system.

Navigational Message

L1OC

Full-length String for L1OC Navigational Message			
Field		**Size, bits**	**Description**
Timecode	CMB	12	Constant bit sequence 0101 1111 0001 (5F1h)
String type	Тип	6	Type of the navigational message
Satellite ID	j	6	System ID number of the satellite (1 to 63; 0 is reserved until FDMA signal switch-off)
Satellite state	Гʲ	1	This satellite is: 0 — healthy, 1 — in error state
Data reliability	lʲ	1	Transmitted navigational messages are: 0 — valid, 1 — unreliable

Ground control callback	П1	4	(Reserved for system use)
Orientation mode	П2	1	Satellite orientiation mode is: 0 — Sun sensor control, 1 — executing predictive thrust or mode transition
UTC correction	KP	2	On the last day of the current quarter, at 00:00 (24:00), a UTC leap second is: 0 — not expected, 1 — expected with positive value, 2 — unknown, 3 — expected with negative value
Execute correction	A	1	After the end of the current string, UTC correction is: 0 — not expected, 1 — expected
Satellite time	OMB	16	Onboard time of the day in 2 s intervals (0 to 43199)
Information		184	Content of the information field is defined by string type
CRC	ЦК	16	Cyclic redundancy code
Total		250	

L3OC

Full-length String for L3OC Navigation Message			
Field		**Size, bits**	**Description**
Timecode	CMB	20	Constant bit sequence 0000 0100 1001 0100 1110 (0494Eh)
String type	Тип	6	Type of the navigational message
Satellite time	OMB	15	Onboard time of the day in 3 s intervals (0 to 28799)
Satellite ID	j	6	
Satellite state	Гj	1	
Data reliability	lj	1	
Ground control callback	П1	4	The same as in L1OC signal
Orientation mode	П2	1	
UTC correction	KP	2	
Execute correction	A	1	
Information		219	Content of the information field is defined by string type
CRC	ЦК	24	Cyclic redundancy code
Total		300	

Common Properties of Open CDMA Signals

String Types for Navigational Signals	
Type	**Content of the information field**
0	(Reserved for system use)

1	Short string for the negative leap second
2	Long string for the positive leap second
10, 11, 12	Real-time information (ephemerides and time-frequency offsets). Transmitted as a packet of three strings in sequence
16	Satellite orientation parameters for the predictive thrust maneuver
20	Almanac
25	Earth rotation parameters, ionosphere models, and time scale model for the difference between UTC(SU) and TAI
31, 32	Parameters of long-term movement model
50	Cospas-Sarsat service message — L1OC signal only
60	Text message

Information Field of a String Type 20 (almanac) for The Orbit Type 0			
Field		**Size, bits**	**Weight of the low bit**	**Description**
Orbit type	TO	2	1	0 — curcular orbit with 19100 km altitude
Satellite number	N_S	6	1	Total number of satellites transmitting CDMA signals (1 to 63) which are referenced to in the almanac
Almanac age	E_A	6	1	Number of full days passed since the last almanac update
Current day	N_A	11	1	Day number (1 to 1461) within a four-year interval starting on January 1 of the last leap year according to Moscow decree time
Signal status	PC_A	5	1	Bit field encoding types of CDMA signals transmitted by the satellite. Three highest bits correspond to signals L1, L2 и L3: 0 — transmitted, 1 — not transmitted
Satellite type	PC_A	3	1	Satellite model and the set of transmitted CDMA signals: 0 — Glonass-M (L3 signal), 1 — Glonass-K1 (L3 signal), 2 — Glonass-K1 (L2 and L3 signals), 3 — Glonass-K2 (L1, L2, and L3 signals)
Time correction	τ_A	14	2^{-20}	Rough correction from onboard time scale to the GLONASS time scale ($\pm 7.8 \times 10^{-3}$ c)
Ascension	λ_A	21	2^{-20}	Longitude of the satellite's first orbital node (± 1 half-cycles)
Ascension time	$t_{\lambda A}$	21	2^{-5}	Time of the day when the satellite is crossing its first orbital node (0 to 44100 s)
Inclination	Δi_A	15	2^{-20}	Adjustments to nominal inclination (64,8°) of the satellite orbit at the moment of ascension (± 0.0156 half-cycles)
Eccentricity	ε_A	15	2^{-20}	Eccentricity of the satellite orbit at the ascension time (0 to 0.03)
Perigee	ω_A	16	2^{-15}	Argument to satellite's perigee at the ascension time (± 1 half-cycles)
Period	ΔT_A	19	2^{-9}	Adjustments to the satellite's nominal draconic orbital period (40544 s) at the moment of ascension (± 512 s)
Period change	$\Delta \dot{T}_A$	7	2^{-14}	Speed of change of the draconic orbital period at the moment of ascension ($\pm 3.9 \times 10^{-3}$ s/orbit)
(Reserved)		L1OC: 23	-	
		L3OC: 58		

- Navigational message field j (satellite ID) references the satellite for the transmitted almanac (jA).

- The set of almanac parameters depends on the orbit type. Satellites with geosynchronous, medium-Earth, and high-elliptical orbits could be employed in the future.

- In a departure from the Gregorian calendar, all years exactly divisible by 100 (i.e. 2100 and so on) are treated as leap years.

Satellites

The main contractor of the GLONASS program is Joint Stock Company Reshetnev Information Satellite Systems (ISS Reshetnev, formerly called NPO-PM). The company, located in Zheleznogorsk, is the designer of all GLONASS satellites, in cooperation with the Institute for Space Device Engineering and the Russian Institute of Radio Navigation and Time. Serial production of the satellites is accomplished by the company PC Polyot in Omsk.

Over the three decades of development, the satellite designs have gone through numerous improvements, and can be divided into three generations: the original GLONASS (since 1982), GLONASS-M (since 2003) and GLONASS-K (since 2011). Each GLONASS satellite has a GRAU designation 11F654, and each of them also has the military "Cosmos-NNNN" designation.

First Generation

The true first generation of GLONASS (also called Uragan) satellites were all three-axis stabilized vehicles, generally weighing 1,250 kilograms (2,760 lb) and were equipped with a modest propulsion system to permit relocation within the constellation. Over time they were upgraded to Block IIa, IIb, and IIv vehicles, with each block containing evolutionary improvements.

Six Block IIa satellites were launched in 1985–1986 with improved time and frequency standards over the prototypes, and increased frequency stability. These spacecraft also demonstrated a 16-month average operational lifetime. Block IIb spacecraft, with a two-year design lifetimes, appeared in 1987, of which a total of 12 were launched, but half were lost in launch vehicle accidents. The six spacecraft that made it to orbit worked well, operating for an average of nearly 22 months.

Block IIv was the most prolific of the first generation. Used exclusively from 1988 to 2000, and continued to be included in launches through 2005, a total of 25 satellites were launched. The design life was three years, however numerous spacecraft exceeded this, with one late model lasting 68 months, nearly double.

Block II satellites were typically launched three at a time from the Baikonur Cosmodrome using Proton-K Blok-DM-2 or Proton-K Briz-M boosters. The only exception was when, on two launches, an Etalon geodetic reflector satellite was substituted for a GLONASS satellite.

Second Generation

The second generation of satellites, known as Glonass-M, were developed beginning in 1990 and first launched in 2003. These satellites possess a substantially increased lifetime of seven years and weigh slightly more at 1,480 kilograms (3,260 lb). They are approximately 2.4 m (7 ft 10 in) in diameter and 3.7 m (12 ft) high, with a solar array span of 7.2 m (24 ft) for an electrical power generation capability of 1600 watts at launch. The aft payload structure houses 12 primary antennas for L-band transmissions. Laser corner-cube reflectors are also carried to aid in precise orbit determination and geodetic research. On-board cesium clocks provide the local clock source. Glonass-M consisting 31 satellites ranging from satellite index 21 - 92 and with 4 spare active satellites.

A total of 41 second generation satellites were launched through the end of 2013. As with the previous generation, the second generation spacecraft were launched in triplets using Proton-K Blok-DM-2 or Proton-K Briz-M boosters. Some where launched alone with Soyuz-2-1b/Fregat

On July 30, 2015, ISS Reshetnev announced that it had completed the last GLONASS-M (N° 61) spacecraft and it was putting it in storage waiting for launch, along an additional eight already built satellites.

Third Generation

GLONASS-K is a substantial improvement of the previous generation: it is the first unpressurised GLONASS satellite with a much reduced mass (750 kilograms (1,650 lb) versus 1,450 kilograms (3,200 lb) of GLONASS-M). It has an operational lifetime of 10 years, compared to the 7-year lifetime of the second generation GLONASS-M. It will transmit more navigation signals to improve the system's accuracy—including new CDMA signals in the L3 and L5 bands, which will use modulation similar to modernized GPS, Galileo, and Compass. Glonass-K consist of 26 satellites having satellite index 65-98 and widely used in Russian Military space. The new satellite's advanced equipment—made solely from Russian components—will allow the doubling of GLONASS' accuracy. As with the previous satellites, these are 3-axis stabilized, nadir pointing with dual solar arrays. The first GLONASS-K satellite was successfully launched on 26 February 2011.

Due to their weight reduction, GLONASS-K spacecraft can be launched in pairs from the Plesetsk Cosmodrome launch site using the substantially lower cost Soyuz-2.1b boosters or in six-at-once from the Baikonur Cosmodrome using Proton-K Briz-M launch vehicles.

Ground Control

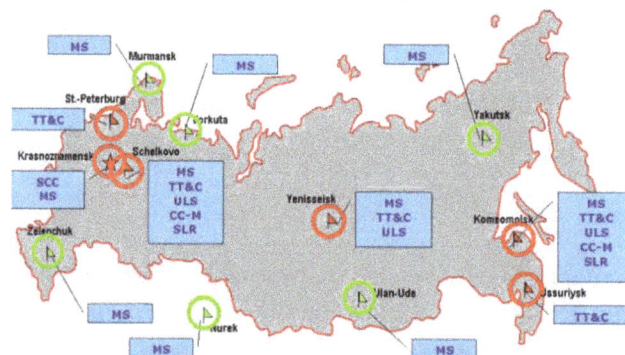

Map depicting ground control stations

The ground control segment of GLONASS is almost entirely located within former Soviet Union territory, except for several in Brazil.

The GLONASS ground segment consists of:

- a system control centre;

- five Telemetry, Tracking and Command centers;

- two Laser Ranging Stations; and

- ten Monitoring and Measuring Stations.

Location	System control	Telemetry, Track-ing and Command	Central clock	Upload stations	Laser Ranging	Monitoring and Measuring
Kras-noznamensk	x	-	-	-	-	x
Schelkovo	-	x	x	x	x	x
Komsomoisk	-	x	-	x	x	x
St-Peteburg	-	x	-	-	-	-
Ussuriysk	-	x	-	-	-	-
Yenisseisk	-	x	-	x	-	x
Yakutsk	-	-	-	-	-	x
Ulan-Ude	-	-	-	-	-	x
Nurek	-	-	-	-	-	x
Vorkuta	-	-	-	-	-	x
Murmansk	-	-	-	-	-	x
Zelenchuk	-	-	-	-	-	x

Receivers

Septentrio, Topcon, C-Nav, JAVAD, Magellan Navigation, Novatel, Leica Geosystems, Hemi-sphere GNSS and Trimble Inc produce GNSS receivers making use of GLONASS. NPO Progress describes a receiver called *GALS-A1*, which combines GPS and GLONASS reception. SkyWave Mobile Communications manufactures an Inmarsat-based satellite communications terminal that uses both GLONASS and GPS. As of 2011, some of the latest receivers in the Garmin eTrex line also support GLONASS (along with GPS). Garmin also produce a standalone Bluetooth receiver, the GLO for Aviation, which combines GPS, WAAS and GLONASS. Various smartphones from 2011 onwards have integrated GLONASS capability, including devices from Xiaomi Tech Company (Xiaomi Phone 2), Sony Ericsson, ZTE, Huawei, Samsung (Galaxy Note, Galaxy Note II, Galaxy S3, Galaxy S4), Apple (iPhone 4S, iPhone 5, iPhone 5C, iPhone 5S, iPhone 6 and iPhone 6 Plus, iPhone 6s, iPhone 6s Plus, iPhone SE, iPhone 7 and iPhone 7 Plus), iPad Mini (LTE models only), iPad Mini 2 (LTE models only), iPad Mini 3 (LTE models only), iPad Mini 4 (LTE models only) iPad (3rd generation and 4th Generation, 4G and LTE models only [respectively]), iPad Air (LTE models only) and iPad Air 2 (LTE models only) and Apple's flagship iPad Pro 12.9" and 9.7", HTC, LG, Motorola and Nokia.

Status

Availability

As of 18 March 2017, the GLONASS constellation status is:

Total	27 SC
Operational	24 SC (Glonass-M/K)
In commissioning	0 SC
In maintenance	0 SC
Under check by the Satellite Prime Contractor	1 SC
Spares	1 SC (Glonass-M)
In flight tests phase	1 SC (Glonass-K)
	–

The system requires 18 satellites for continuous navigation services covering the entire territory of the Russian Federation, and 24 satellites to provide services worldwide. The GLONASS system covers 100% of worldwide territory.

On 2 April 2014 the system experienced a technical failure that resulted in practical unavailability of the navigation signal for around 12 hours.

On 14–15 April 2014 nine GLONASS satellites experienced a technical failure due to software problems.

On 19 February 2016 three GLONASS satellites experienced a technical failure: the batteries of GLONASS-738 exploded, the batteries of GLONASS-737 were depleted, and GLONASS-736 experienced a stationkeeping failure due to human error during maneuvering. GLONASS-737 and GLONASS-736 are expected to be operational again after maintenance, and one new satellite (GLONASS-751) to replace GLONASS-738 is expected to complete commissioning in early March. The full capacity of the satellite group is expected to be restored in the middle of March. After the launching of two new satellites and maintenance of two others, the full capacity of the satellite group was restored.

Accuracy

According to Russian System of Differential Correction and Monitoring's data, as of 2010, precision of GLONASS navigation definitions (for p=0.95) for latitude and longitude were 4.46–7.38 metres (14.6–24.2 ft) with mean number of navigation space vehicles (NSV) equals 7—8 (depending on station). In comparison, the same time precision of GPS navigation definitions were 2.00–8.76 metres (6 ft 7 in–28 ft 9 in) with mean number of NSV equals 6—11 (depending on station). Civilian GLONASS used alone is therefore very slightly less accurate than GPS. On high latitudes (north or south), GLONASS' accuracy is better than that of GPS due to the orbital position of the satellites.

Some modern receivers are able to use both GLONASS and GPS satellites together, providing greatly improved coverage in urban canyons and giving a very fast time to fix due to over 50 satellites being available. In indoor, urban canyon or mountainous areas, accuracy can be greatly improved over using GPS alone. For using both navigation systems simultaneously, precision of GLONASS/GPS navigation definitions were 2.37–4.65 metres (7 ft 9 in–15 ft 3 in) with mean number of NSV equals 14—19 (depends on station).

In May 2009, Anatoly Perminov, then director of the Russian Federal Space Agency, stated that actions were undertaken to expand GLONASS's constellation and to improve the ground segment to increase the navigation definition of GLONASS to an accuracy of 2.8 metres (9 ft 2 in) by 2011. In particular, the latest satellite design, GLONASS-K has the ability to double the system's accuracy once introduced. The system's ground segment is also to undergo improvements. As of early 2012, sixteen positioning ground stations are under construction in Russia and in the Antarctic at the Bellingshausen and Novolazarevskaya bases. New stations will be built around the southern hemisphere from Brazil to Indonesia. Together, these improvements are expected to bring GLONASS' accuracy to 0.6 m or better by 2020.

Inception and Design

A GLONASS satellite

The first satellite-based radio navigation system developed in the Soviet Union was Tsiklon, which had the purpose of providing ballistic missile submarines a method for accurate positioning. 31 Tsiklon satellites were launched between 1967 and 1978. The main problem with the system was that, although highly accurate for stationary or slow-moving ships, it required several hours of observation by the receiving station to fix a position, making it unusable for many navigation purposes and for the guidance of the new generation of ballistic missiles. In 1968–1969, a new navigation system, which would support not only the navy, but also the air, land and space forces, was conceived. Formal requirements were completed in 1970; in 1976, the government made a decision to launch development of the "Unified Space Navigation System GLONASS".

The task of designing GLONASS was given to a group of young specialists at NPO PM in the city of Krasnoyarsk-26 (today called Zheleznogorsk). Under the leadership of Vladimir Cheremisin, they developed different proposals, from which the institute's director Grigory Chernyavsky selected the final one. The work was completed in the late 1970s; the system consists of 24 satellites operating at an altitude of 20,000 kilometres (12,000 mi) in medium circular orbit. It would be able to promptly fix the receiving station's position based on signals from four satellites, and also reveal the object's speed and direction. The satellites would be launched three at a time on the heavy-lift Proton rocket. Due to the large number of satellites needed for the program, NPO PM delegated the manufacturing of the satellites to PO Polyot in Omsk, which had better production capabilities.

Originally, GLONASS was designed to have an accuracy of 65 metres (213 ft), but in reality it had an accuracy of 20 metres (66 ft) in the civilian signal and 10 metres (33 ft) in the military signal. The first generation GLONASS satellites were 7.8 metres (26 ft) tall, had a width of 7.2 metres (24 ft), measured across their solar panels, and a mass of 1,260 kilograms (2,780 lb).

Achieving Full Orbital Constellation

In the early 1980s, NPO PM received the first prototype satellites from PO Polyot for ground tests. Many of the produced parts were of low quality and NPO PM engineers had to perform substantial redesigning, leading to a delay. On 12 October 1982, three satellites, designated Kosmos-1413, Kosmos-1414, and Kosmos-1415 were launched aboard a Proton rocket. As only one GLONASS satellite was ready in time for the launch instead of the expected three, it was decided to launch it along with two mock-ups. The USA media reported the event as a launch of one satellite and "two secret objects." For a long time, the USA could not find out the nature of those "objects". The Telegraph Agency of the Soviet Union (TASS) covered the launch, describing GLONASS as a system "created to determine positioning of civil aviation aircraft, navy transport and fishing-boats of the Soviet Union".

From 1982 to April 1991, the Soviet Union successfully launched a total of 43 GLONASS-related satellites plus five test satellites. When the Soviet Union disintegrated in 1991, twelve GLONASS satellites in two planes were operational; enough to allow limited use of the system (to cover the entire territory of the Union, 18 satellites would have been necessary.) The Russian Federation took over control of the constellation and continued its development. In 1993, the system, now consisting of 12 satellites, was formally declared operational and in December 1995 it was brought to a fully operational constellation of 24 satellites. This brought the precision of GLONASS on a par with the USA GPS system, which had achieved full operation a year earlier.

Economic Crisis

Since the first generation satellites operated for three years each, to keep the system at full capacity, two launches per year would have been necessary to maintain the full network of 24 satellites. However, in the financially difficult period of 1989–1999, the space program's funding was cut by 80% and Russia consequently found itself unable to afford this launch rate. After the full complement was achieved in December 1995, there were no further launches until December 1999. As a result, the constellation reached its lowest point of just six operational satellites in 2001. As a prelude to demilitarisation, responsibility of the program was transferred from the Ministry of Defence to Russia's civilian space agency Roscosmos.

Renewed Efforts and Modernization

President Vladimir Putin with a GLONASS car navigation device. As President, Putin paid special attention to the development of GLONASS.

In the 2000s, during Vladimir Putin's presidency, the Russian economy recovered and state finances improved considerably. Putin himself took special interest in GLONASS and the system's restoration was made one of the government's top priorities. For this purpose, on August 2001, the Federal Targeted Program "Global Navigation System" 2002–2011 (Government Decision No. 587) was launched. The program was given a budget of $420 million and aimed at restoring the full constellation by 2009.

On 10 December 2003, the second generation satellite design, GLONASS-M, was launched for the first time. It had a slightly larger mass than the baseline GLONASS, standing at 1,415 kilograms (3,120 lb), but it had seven years lifetime, four years longer than the lifetime of the original GLONASS satellite, decreasing the required replacement rate. The new satellite also had better accuracy and ability to broadcast two extra civilian signals.

In 2006, Defence Minister Sergey Ivanov ordered one of the signals (with an accuracy of 30 metres (98 ft)) to be made available to civilian users. Putin, however, was not satisfied with this, and demanded that the whole system should be made fully available to everyone. Consequently, on 18 May 2007, all restrictions were lifted. The accurate, formerly military-only signal with a precision of 10 metres (33 ft), has since then been freely available to civilian users.

During the middle of the first decade of the 21st century, the Russian economy boomed, resulting in substantial increases in the country's space budget. In 2007, the financing of the GLONASS program was increased considerably; its budget was more than doubled. While in 2006 the GLONASS had received $181 million from the federal budget, in 2007 the amount was increased to $380 million.

In the end, 140.1 billion rubles ($4.7 billion) were spent on the program 2001–2011, making it Roscosmos' largest project and consuming a third of its 2010 budget of 84.5 billion rubles.

For the period of 2012 to 2020 320 billion rubles ($10 billion) were allocated to support the system.

Restoring Full Capacity

In June 2008, the system consisted of 16 satellites, 12 of which were fully operational at the time. At this point, Roscosmos aimed at having a full constellation of 24 satellites in orbit by 2010, one year later than previously planned.

In September 2008, Prime Minister Vladimir Putin signed a decree allocating additional 67 billion rubles ($2.6 billion) to GLONASS from the federal budget.

Promoting Commercial Use

Although the GLONASS constellation has reached global coverage, its commercialisation, especially development of the user segment, has been lacking compared to the American GPS. For example, the first commercial Russian-made GLONASS navigation device for cars, Glospace SGK-70, was introduced in 2007, but it was much bigger and costlier than similar GPS receivers. In late 2010, there were only a handful of GLONASS receivers on the market, and few of them were meant for ordinary consumers. To improve the situation, the Russian government has been actively promoting GLONASS for civilian use.

To improve development of the user segment, on 11 August 2010, Sergei Ivanov announced a plan to introduce a 25% import duty on all GPS-capable devices, including mobile phones, unless they are compatible with GLONASS. The government also planned to force all car manufacturers in Russia to support GLONASS starting from 2011. This would affect all car makers, including foreign brands like Ford and Toyota, which have car assembly facilities in Russia.

GPS and phone baseband chips from major vendors Qualcomm, Exynos and Broadcom all support GLONASS in combination with GPS.

In April 2011, Sweden's Swepos—a national network of satellite reference stations that provides real-time positioning data with meter accuracy—became the first known foreign company to use GLONASS.

Smartphones and Tablets also saw implementation of GLONASS support in 2011 with devices released that year from Xiaomi Tech Company (Xiaomi Phone 2), Sony Ericsson, Samsung (Galaxy Note, Galaxy Note II, Galaxy SII, Galaxy SIII mini, the Google Nexus 10 in late 2012), Asus, Apple (iPhone 4S and iPad Mini in late 2012) and HTC adding support for the system allowing increased accuracy and lock on speed in difficult conditions.

Finishing the Constellation

Russia's aim of finishing the constellation in 2010 suffered a setback when a December 2010 launch of three GLONASS-M satellites failed. The Proton-M rocket itself performed flawlessly, but the upper stage Blok DM3 (a new version that was to make its maiden flight) was loaded with too much fuel due to a sensor failure. As a result, the upper stage and the three satellites crashed into the Pacific Ocean. Kommersant estimated that the launch failure cost up to $160 million. Russian President Dmitry Medvedev ordered a full audit of the entire program and an investigation into the failure.

Following the mishap, Roscosmos activated two reserve satellites and decided to make the first improved GLONASS-K satellite, to be launched in February 2011, part of the operational constellation instead of mainly for testing as was originally planned. This would bring the total number of satellites to 23, obtaining almost complete worldwide coverage. The GLONASS-K2 was originally scheduled to be launched by 2013, however by 2012 was not expected to be launched until 2015.

In 2010, President Dmitry Medvedev ordered the government to prepare a new federal targeted program for GLONASS, covering the years 2012–2020. The original 2001 program is scheduled to end in 2011. On 22 June 2011, Roscosmos revealed that the agency was looking for a funding of 402 billion rubles ($14.35 billion) for the program. The funds would be spent on maintaining the satellite constellation, on developing and maintaining navigational maps as well as on sponsoring supplemental technologies to make GLONASS more attractive to users.

On 2 October 2011 the 24th satellite of the system, a GLONASS-M, was successfully launched from Plesetsk Cosmodrome and is now in service. This made the GLONASS constellation fully restored, for the first time since 1996.

On 5 November 2011 the Proton-M booster successfully put three GLONASS-M units in final orbit.

On Monday 28 November 2011, a Soyuz rocket, launched from the Plesetsk Cosmodrome Space Centre, placed a single GLONASS-M satellite into orbit into Plane 3.

On 26 April 2013 a single GLONASS-M satellite was delivered to the orbit by Soyuz rocket from Plesetsk Cosmodrome, restoring the constellation to 24 operational satellites, the minimum to provide global coverage.

On 2 July 2013 a Proton-M rocket, carrying 3 GLONASS-M satellites, crashed during takeoff from Baikonur Cosmodrome. It veered off the course just after leaving the pad and plunged into the ground nose first. The rocket employed a DM-03 booster, for the first time since the December 2010 launch, when the vehicle had also failed, resulting in a loss of another 3 satellites.

However, as of 2014, while the system was completed from technical point of view, the operational side was still not closed by the Ministry of Defense and its formal status was still "in development".

On 7 December 2015, the system was officially completed.

Galileo (Satellite Navigation)

Galileo is the global navigation satellite system (GNSS) that is currently being created by the European Union (EU) through the European Space Agency (ESA) and the European GNSS Agency (GSA), headquartered in Prague in the Czech Republic, with two ground operations centres, Oberpfaffenhofen near Munich in Germany and Fucino in Italy. The €5 billion project is named after the Italian astronomer Galileo Galilei. One of the aims of Galileo is to provide an independent high-precision positioning system so European nations do not have to rely on the Russian GLONASS, Chinese BeiDou or US GPS systems, which could be disabled or degraded by their operators at any time. The use of basic (lower-precision) Galileo services will be free and open to everyone. The higher-precision capabilities will be available for paying commercial users. Galileo is intended to provide horizontal and vertical position measurements within 1-metre precision, and better positioning services at high latitudes than other positioning systems.

Galileo is to provide a new global search and rescue (SAR) function as part of the MEOSAR system. Satellites will be equipped with a transponder which will relay distress signals from emergency beacons to the Rescue coordination centre, which will then initiate a rescue operation. At the same time, the system is projected to provide a signal, the Return Link Message (RLM), *to* the emergency beacon, informing them that their situation has been detected and help is on the way. This latter feature is new and is considered a major upgrade compared to the existing Cospas-Sarsat system, which does not provide feedback to the user. Tests in February 2014 found that for Galileo's search and rescue function, operating as part of the existing International Cospas-Sarsat Programme, 77% of simulated distress locations can be pinpointed within 2 km, and 95% within 5 km.

The first Galileo test satellite, the GIOVE-A, was launched 28 December 2005, while the first satellite to be part of the operational system was launched on 21 October 2011. As of December 2016 the system has 18 of 30 satellites in orbit. Galileo started offering Early Operational Capability (EOC) on 15 December 2016 and is expected to reach Full Operational Capability (FOC) in 2019. The complete 30-satellite Galileo system (24 operational and 6 active spares) is expected by 2020.

History

Headquarters of the Galileo system in Prague

Main Objectives

In 1999, the different concepts of the three main contributors of ESA (Germany, France and Italy) for Galileo were compared and reduced to one by a joint team of engineers from all three countries. The first stage of the Galileo programme was agreed upon officially on 26 May 2003 by the European Union and the European Space Agency. The system is intended primarily for civilian use, unlike the more military-orientated systems of the United States (GPS), Russia (GLONASS), and China (Beidou-1/2, COMPASS). The European system will only be subject to shutdown for military purposes in extreme circumstances (like armed conflict). It will be available at its full precision to both civil and military users.

Funding

The European Commission had some difficulty funding the project's next stage, after several allegedly "per annum" sales projection graphs for the project were exposed in November 2001 as "cumulative" projections which for each year projected included all previous years of sales. The attention that was brought to this multibillion-euro growing error in sales forecasts resulted in a general awareness in the Commission and elsewhere that it was unlikely that the program would yield the return on investment that had previously been suggested to investors and decision-makers. On 17 January 2002, a spokesman for the project stated that, as a result of US pressure and economic difficulties, "Galileo is almost dead."

A few months later, however, the situation changed dramatically. European Union member states decided it was important to have a satellite-based positioning and timing infrastructure that the US could not easily turn off in times of political conflict.

The European Union and the European Space Agency agreed in March 2002 to fund the project, pending a review in 2003 (which was completed on 26 May 2003). The starting cost for the period ending in 2005 is estimated at €1.1 billion. The required satellites (the planned number is 30) were to be launched between 2011 and 2014, with the system up and running and under civilian control from 2019. The final cost is estimated at €3 billion, including the infrastructure on Earth, constructed in 2006 and 2007. The plan was for private companies and investors to invest at least two-thirds of the cost of implementation, with the EU and ESA dividing the remaining cost. The base *Open Service* is to be available without charge to anyone with a Galileo-compatible receiver, with an encrypted higher-bandwidth improved-precision *Commercial Service* available at a cost. By early 2011 costs for the project had run 50% over initial estimates.

The German Aerospace Center (DLR) contributes the largest portion of the Galileo funds, and is crucial in the development and application of the system with its facilities of the Earth Observation Center, and the Institute for Communication and Navigation in Neustrelitz.

Tension with the United States

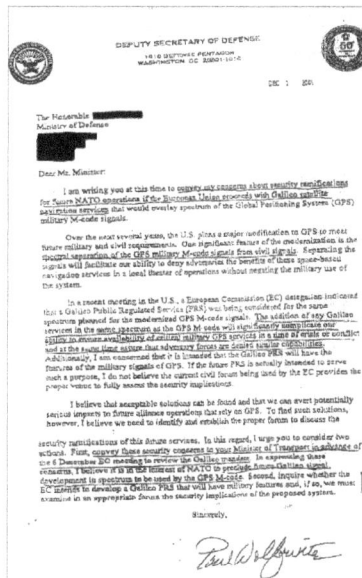

A December 2001 letter from U.S. Deputy Secretary of Defense Paul Wolfowitz to the Ministers of the EU states as part of the US-lobbying campaign against Galileo

Galileo is intended to be an EU civilian GNSS that allows all users access to it. Initially GPS reserved the highest quality signal for military use, and the signal available for civilian use was intentionally degraded (Selective Availability). This changed with President Bill Clinton signing a policy directive in 1996 to turn off Selective Availability. Since May 2000 the same precision signal has been provided to both civilians and the military.

Since Galileo was designed to provide the highest possible precision (greater than GPS) to anyone, the US was concerned that an enemy could use Galileo signals in military strikes against the US and its allies (some weapons like missiles use GNSSs for guidance). The frequency initially chosen for Galileo would have made it impossible for the US to block the Galileo signals without also interfering with its own GPS signals. The US did not want to lose their GNSS capability with GPS while denying enemies the use of GNSS. Some US officials became especially concerned when Chinese interest in Galileo was reported.

An anonymous EU official claimed that the US officials implied that they might consider shooting down Galileo satellites in the event of a major conflict in which Galileo was used in attacks against American forces. The EU's stance is that Galileo is a neutral technology, available to all countries and everyone. At first, EU officials did not want to change their original plans for Galileo, but have since reached a compromise, that Galileo was to use a different frequency. This allowed the blocking or jamming of either GNSS without affecting the other (jam Galileo without affecting GPS, or jam GPS but not Galileo), giving the US a greater advantage in conflicts in which it has the electronic warfare upper hand.

GPS and Galileo

One of the reasons given for developing Galileo as an independent system was that position information from GPS can be made significantly inaccurate by the deliberate application of universal Selective Availability (SA) by the US military. GPS is widely used worldwide for civilian applications; Galileo's proponents argued that civil infrastructure, including airplane navigation and landing, should not rely solely upon a system with this vulnerability.

On 2 May 2000, SA was disabled by the President of the United States, Bill Clinton; in late 2001 the entity managing the GPS confirmed that they did not intend to enable selective availability ever again. Though Selective Availability capability still exists, on 19 September 2007 the US Department of Defense announced that newer GPS satellites would not be capable of implementing Selective Availability; the wave of Block IIF satellites launched in 2009, and all subsequent GPS satellites, are stated not to support SA. As old satellites are replaced in the GPS Block IIIA program, SA will cease to be an option. The modernisation programme also contains standardised features that allow GPS III and Galileo systems to inter-operate, allowing receivers to be developed to utilise GPS and Galileo together to create an even more precise GNSS.

Cooperation with the United States

In June 2004, in a signed agreement with the United States, the European Union agreed to switch to a modulation known as BOC(1,1) (Binary Offset Carrier 1.1) allowing the coexistence of both GPS and Galileo, and the future combined use of both systems.

The European Union also agreed to address the "mutual concerns related to the protection of allied and US national security capabilities."

First Experimental Satellites: GIOVE-A and GIOVE-B

The first experimental satellite, GIOVE-A, was launched in December 2005 and was followed by a second test satellite, GIOVE-B, launched in April 2008. After successful completion of the In-Orbit Validation (IOV) phase, additional satellites were launched. On 30 November 2007 the 27 EU transport ministers involved reached an agreement that Galileo should be operational by 2013, but later press releases suggest it was delayed to 2014.

Funding Again, Governance Issues

In mid-2006 the public/private partnership fell apart, and the European Commission decided to nationalise the Galileo programme.

In early 2007 the EU had yet to decide how to pay for the system and the project was said to be "in deep crisis" due to lack of more public funds. German Transport Minister Wolfgang Tiefensee was particularly doubtful about the consortium's ability to end the infighting at a time when only one testbed satellite had been successfully launched.

Although a decision was yet to be reached, on 13 July 2007 EU countries discussed cutting €548m ($755m, £370m) from the union's competitiveness budget for the following year and shifting some of these funds to other parts of the financing pot, a move that could meet part of the cost of the union's Galileo satellite navigation system. European Union research and development projects could be scrapped to overcome a funding shortfall.

In November 2007, it was agreed to reallocate funds from the EU's agriculture and administration budgets and to soften the tendering process in order to invite more EU companies.

In April 2008, the EU transport ministers approved the Galileo Implementation Regulation. This allowed the €3.4bn to be released from the EU's agriculture and administration budgets to allow the issuing of contracts to start construction of the ground station and the satellites.

In June 2009, the European Court of Auditors published a report, pointing out governance issues, substantial delays and budget overruns that led to project stalling in 2007, leading to further delays and failures.

In October 2009, the European Commission cut the number of satellites definitively planned from 28 to 22, with plans to order the remaining six at a later time. It also announced that the first OS, PRS and SoL signal would be available in 2013, and the CS and SOL some time later. The €3.4 billion budget for the 2006–2013 period was considered insufficient. In 2010 the think-tank Open Europe estimated the total cost of Galileo from start to 20 years after completion at €22.2 billion, borne entirely by taxpayers. Under the original estimates made in 2000, this cost would have been €7.7 billion, with €2.6 billion borne by taxpayers and the rest by private investors.

In November 2009, a ground station for Galileo was inaugurated near Kourou (French Guiana).

The launch of the first four in-orbit validation (IOV) satellites was planned for the second half of 2011, and the launch of full operational capability (FOC) satellites was planned to start in late 2012.

In March 2010 it was verified that the budget for Galileo would only be available to provide the 4 IOV and 14 FOC satellites by 2014, with no funds then committed to bring the constellation above this 60% capacity. Paul Verhoef, the satellite navigation program manager at the European Commission, indicated that this limited funding would have serious consequences commenting at one point "To give you an idea, that would mean that for three weeks in the year you will not have satellite navigation" in reference to the proposed 18-vehicle constellation.

In July 2010, the European Commission estimated further delays and additional costs of the project to grow up to €1.5-€1.7 billion, and moved the estimated date of completion to 2018. After completion the system will need to be subsidised by governments at €750 million per year. An additional €1.9 billion was planned to be spent bringing the system up to the full complement of 30 satellites (27 operational + 3 active spares).

In December 2010, EU ministers in Brussels voted Prague, in the Czech Republic, as the headquarters of the Galileo project.

In January 2011, infrastructure costs up to 2020 were estimated at €5.3 billion. In that same month, Wikileaks revealed that Berry Smutny, the CEO of the German satellite company OHB-System, said that Galileo "is a stupid idea that primarily serves French interests". The BBC understood in 2011 that €500 million (£440M) would become available to make the extra purchase, taking Galileo within a few years from 18 operational satellites to 24.

Galileo launch on a Soyuz rocket, 21 October 2011

The first two Galileo In-Orbit Validation satellites were launched by Soyuz ST-B flown from Guiana Space Centre on 21 October 2011, and the remaining two on 12 October 2012.

Twenty-two further satellites with Full Operational Capability (FOC) were on order as of 2012. The first four pairs of satellites were launched on 22 August 2014, 27 March 2015, 11 September 2015 and 17 December 2015.

Clock Failures

In January 2017, news agencies reported that six of the passive hydrogen maser and three of the rubidium atomic clocks had failed. Four of the full operational satellites have each lost at least one clock; but no satellite has lost more than two. The operation of the constellation has not been affected as each satellite is launched with three spare clocks. The possibility of a systematic flaw is being considered. The Swiss producer of both onboard clocktypes SpectraTime declined to comment. According to ESA they concluded with their industrial partners for the rubidium atomic clocks some implemented testing and operational measures were required. Additionally some refurbishment is required for the rubidium atomic clocks that still have to be launched. For the passive hydrogen masers operational measures are being studied to reduce the risk of failure. China and India use the same SpectraTime-built atomic clocks in their satellite navigation systems. ESA has contacted the Indian Space Research Organisation who initially reported not having experienced similar failures. However, at the end of January 2017, Indian news outlets reported that all three clocks aboard the IRNSS-1A satellite (launched in July 2013 with a 10-year life expectancy) had failed and that a replacement satellite would be launched in the second half of 2017.

International Involvement

In September 2003, China joined the Galileo project. China was to invest €230 million (US$302 million, GBP 155 million, CNY 2.34 billion) in the project over the following years.

In July 2004, Israel signed an agreement with the EU to become a partner in the Galileo project.

On 3 June 2005 the EU and Ukraine signed an agreement for Ukraine to join the project, as noted in a press release.

As of November 2005, Morocco also joined the programme.

In Mid-2006, the Public-Private Partnership fell apart and the European Commission decided to nationalise Galileo as an EU programme.

In November 2006, China opted instead to independently develop the Beidou navigation system satellite navigation system.

On 30 November 2007, the 27 member states of the European Union unanimously agreed to move forward with the project, with plans for bases in Germany and Italy. Spain did not approve during the initial vote, but approved it later that day. This greatly improves the viability of the Galileo project: "The EU's executive had previously said that if agreement was not reached by January 2008, the long-troubled project would essentially be dead."

On 3 April 2009, Norway too joined the programme pledging €68.9 million toward development costs and allowing its companies to bid for the construction contracts. Norway, while not a member of the EU, is a member of ESA.

On 18 December 2013, Switzerland signed a cooperation agreement to fully participate in the program, and retroactively contributed €80 million for the period 2008-2013. As a member of ESA, it already collaborated in the development of the Galileo satellites, contributing the state-of-the-art hydrogen-maser clocks. Switzerland's financial commitment for the period 2014-2020 will be calculated in accordance with the standard formula applied for the Swiss participation in the EU research Framework Programme.

System Description

Space Segment

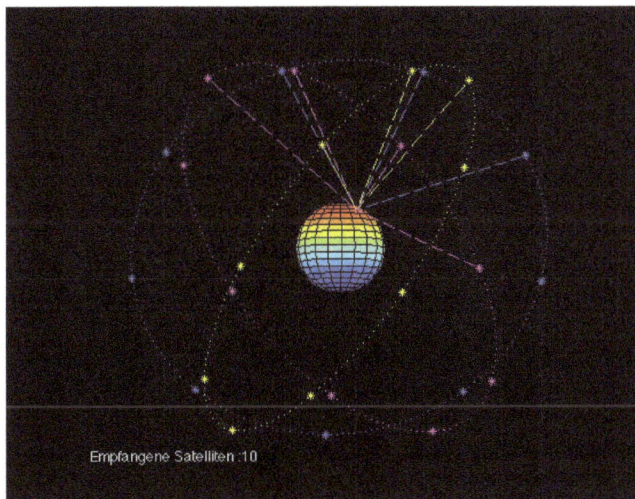

Constellation visibility from a location on Earth's surface

As of 2012, the system is scheduled to reach full operation in 2020 with the following specifications:

- 30 in-orbit spacecraft (24 in full service and 6 spares)

- Orbital altitude: 23,222 km (MEO)

- 3 orbital planes, 56° inclination, ascending nodes separated by 120° longitude (8 operational satellites and 2 active spares per orbital plane)

- Satellite lifetime: >12 years

- Satellite mass: 675 kg

- Satellite body dimensions: 2.7 m × 1.2 m × 1.1 m

- Span of solar arrays: 18.7 m

- Power of solar arrays: 1.5 kW (end of life)

Ground Segment

The system's orbit and signal accuracy is controlled by a ground segment consisting of:

- 2 Ground Control Centre, located in Oberpfaffenhofen and Fucino for Satellite and Mission Control

- 5 telemetry, tracking & control (TT&C) stations, located in Kiruna, Kourou, Noumea, Sainte-Marie, Réunion & Redu

- Several worldwide distributed mission data uplink stations (ULS)

- Several worldwide distributed reference sensor stations (GSS)

- A data dissemination network between all geographically distributed locations

Services

The Galileo system will have five main services:

Open access navigation

> This will be available without charge for use by anyone with appropriate mass-market equipment; simple timing, and positioning down to 1 metre.

Commercial navigation (encrypted)

> High precision to the centimetre; guaranteed service for which service providers will charge fees.

Safety of life navigation

> Open service; for applications where guaranteed precision is essential. Integrity messages will warn of errors.

Public regulated navigation (encrypted)

Continuous availability even if other services are disabled in time of crisis; Government agencies will be main users.

Search and rescue

System will pick up distress beacon locations; feasible to send feedback, e.g. confirming help is on its way.

Other secondary services will also be available.

Concept

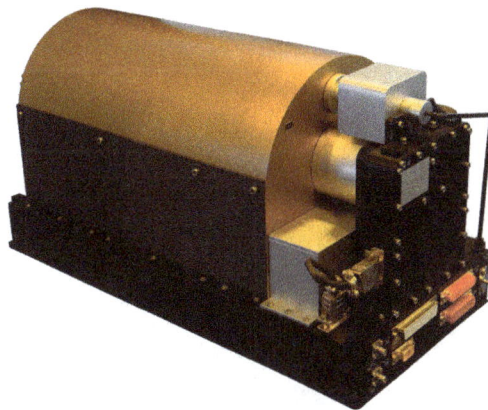

Space Passive Hydrogen Maser used in Galileo satellites as a master clock for an onboard timing system

Each Galileo satellite has two master passive hydrogen maser atomic clocks and two secondary rubidium atomic clocks which are independent of one other. As precise and stable space-qualified atomic clocks are critical components to any satellite-navigation system, the employed quadruple redundancy keeps Galileo functioning when onboard atomic clocks fail in space. The onboard passive hydrogen maser clocks' precision is four times better than the onboard rubidium atomic clocks and estimated at 1 second per 3 million years (a timing error of a nanosecond or 1 billionth of a second (10^{-9} or $^1/_{1,000,000,000}$ s) translates into a 30 cm (11.8 in) positional error on Earth's surface), and will provide an accurate timing signal to allow a receiver to calculate the time that it takes the signal to reach it. The Galileo satellites are configured to run one hydrogen maser clock in primary mode and a rubidium clock as hot backup. Under normal conditions, the operating hydrogen maser clock produces the reference frequency from which the navigation signal is generated. Should the hydrogen maser encounter any problem, an instantaneous switchover to the rubidium clock would be performed. In case of a failure of the primary hydrogen maser the secondary hydrogen maser could be activated by the ground segment to take over within a period of days as part of the redundant system. A clock monitoring and control unit provides the interface between the four clocks and the navigation signal generator unit (NSU). It passes the signal from the active hydrogen master clock to the NSU and also ensures that the frequencies produced by the master clock and the active spare are in phase, so that the spare can take over instantly should the master clock fail. The NSU information is used to calculate the position of the receiver by trilaterating the difference in received signals from multiple satellites.

Constellation

<table>
<tr><td colspan="6" align="center">Summary of Satellites</td></tr>
<tr><td rowspan="2" align="center">Block</td><td rowspan="2" align="center">Launch Period</td><td colspan="3" align="center">Satellite launches</td><td rowspan="2" align="center">Currently in operational orbit and healthy</td></tr>
<tr><td align="center">Full success</td><td align="center">Failure</td><td align="center">Planned</td></tr>
<tr><td align="center">GIOVE</td><td>2005–2008</td><td>2</td><td>0</td><td>0</td><td>0</td></tr>
<tr><td align="center">IOV</td><td>2011–2012</td><td>4</td><td>0</td><td>0</td><td>3</td></tr>
<tr><td align="center">FOC</td><td>From 2014</td><td>12</td><td>2*</td><td>16</td><td>12</td></tr>
<tr><td colspan="2" align="center">Total</td><td>18</td><td>2*</td><td>16</td><td>15</td></tr>
<tr><td colspan="6">* One partial launch failure resulting in 2 satellites orbiting in a degraded orbit</td></tr>
</table>

Galileo Satellite Test Beds: GIOVE

GIOVE-A was successfully launched 28 December 2005

In 2004 the Galileo System Test Bed Version 1 (GSTB-V1) project validated the on-ground algorithms for Orbit Determination and Time Synchronisation (OD&TS). This project, led by ESA and European Satellite Navigation Industries, has provided industry with fundamental knowledge to develop the mission segment of the Galileo positioning system.

- GIOVE-A is the first GIOVE (Galileo In-Orbit Validation Element) test satellite. It was built by Surrey Satellite Technology Ltd (SSTL), and successfully launched on 28 December 2005 by the European Space Agency and the Galileo Joint. Operation of GIOVE-A ensured that Galileo meets the frequency-filing allocation and reservation requirements for the International Telecommunication Union (ITU), a process that was required to be complete by June 2006.

- GIOVE-B, built by Astrium and Thales Alenia Space, has a more advanced payload than GIOVE-A. It was successfully launched on 27 April 2008 at 22:16 UTC (4.16 am Baikonur time) aboard a Soyuz-FG/Fregat rocket provided by Starsem.

A third satellite, GIOVE-A2, was originally planned to be built by SSTL for launch in the second half of 2008. Construction of GIOVE-A2 was terminated due to the successful launch and in-orbit operation of GIOVE-B.

The GIOVE Mission segment operated by European Satellite Navigation Industries used the GIOVE-A/B satellites to provide experimental results based on real data to be used for risk mitigation for the IOV satellites that followed on from the testbeds. ESA organised the global network of ground stations to collect the measurements of GIOVE-A/B with the use of the GETR receivers for further systematic study. GETR receivers are supplied by Septentrio as well as the first Galileo navigation receivers to be used to test the functioning of the system at further stages of its deployment. Signal analysis of GIOVE-A/B data confirmed successful operation of all the Galileo signals with the tracking performance as expected.

In-Orbit Validation (IOV) Satellites

These testbed satellites were followed by four IOV Galileo satellites that are much closer to the final Galileo satellite design. The Search & Rescue feature is also installed. The first two satellites were launched on 21 October 2011 from Guiana Space Centre using a Soyuz launcher, the other two on 12 October 2012. This enables key validation tests, since earth-based receivers such as those in cars and phones need to "see" a minimum of four satellites in order to calculate their position in three dimensions. Those 4 IOV Galileo satellites were constructed by Astrium GmbH and Thales Alenia Space. On 12 March 2013, a first fix was performed using those four IOV satellites. Once this In-Orbit Validation (IOV) phase has been completed, the remaining satellites will be installed to reach the Full Operational Capability.

Full Operational Capability (FOC) Satellites

On 7 January 2010, it was announced that the contract to build the first 14 FOC satellites was awarded to OHB System and Surrey Satellite Technology Limited (SSTL). Fourteen satellites will be built at a cost of €566M (£510M; $811M). Arianespace will launch the satellites for a cost of €397M (£358M; $569M). The European Commission also announced that the €85 million contract for system support covering industrial services required by ESA for integration and validation of the Galileo system had been awarded to Thales Alenia Space. Thales Alenia Space subcontract performances to Astrium GmbH and security to Thales Communications.

In February 2012, an additional order of eight satellites was awarded to OHB Systems for €250M ($327M), after outbidding EADS Astrium tender offer. Thus bringing the total to 22 FOC satellites.

On 7 May 2014, the first two FOC satellites landed in Guyana for their joint launch planned in summer Originally planned for launch during 2013, problems tooling and establishing the production line for assembly led to a delay of a year in serial production of Galileo satellites. These two satellites (Galileo satellites GSAT-201 and GSAT-202) were launched on 22 August 2014. The names of these satellites are Doresa and Milena named after European children who had previously won a drawing contest. On 23 August 2014, launch service provider Arianespace announced that the flight VS09 experienced anomaly and satellites were injected into an incorrect orbit.

Satellites GSAT-203 and GSAT-204 were launched successfully on 27 March 2015 from Guiana Space Centre using a Soyuz four stage launcher. Using the same Soyuz launcher and launchpad, satellites GSAT-205 and GSAT-206 were launched successfully on 11 September 2015.

Satellites GSAT-208 and GSAT-209 were successfully launched from Kourou, French Guiana, using the Soyuz launcher on December 17, 2015.

Satellites GSAT-210 and GSAT-211 were launched on 24 May 2016 and are being commissioned.

Starting in November 2016, deployment of the last twelve satellites will use a modified Ariane 5 launcher, named Ariane 5 ES, capable of placing four Galileo satellites into orbit per launch.

Satellites GSAT-207, GSAT-212, GSAT-213, GSAT-214 were successfully launched from Kourou, French Guiana, on 17 November 2016 on an Ariane 5 ES.

On 15 December 2016, Galileo started offering Initial Operational Capability (IOC). The services currently offered are Open Service, Public Regulated Service and Search and Rescue Service.

Future Evolution

As of 2014, ESA and its industry partners have begun studies on Galileo Second Generation (G2G) satellites, which will be presented to the EC for the 2020s launch period. One idea is to employ electric propulsion, which would eliminate the need for an upper stage during launch and allow satellites from a single batch to be inserted into more than one orbital plane.

Applications and Impact

Science Projects Using Galileo

In July 2006 an international consortium of universities and research institutions embarked on a study of potential scientific applications of the Galileo constellation. This project, named GEO6, is a broad study oriented to the general scientific community, aiming to define and implement new applications of Galileo.

Among the various GNSS users identified by the Galileo Joint Undertaking, the GEO6, project addresses the Scientific User Community (UC).

The GEO6 project aims at fostering possible novel applications within the scientific UC of GNSS signals, and particularly of Galileo.

The AGILE project is an EU-funded project devoted to the study of the technical and commercial aspects of location-based services (LBS). It includes technical analysis of the benefits brought by Galileo (and EGNOS) and studies the hybridisation of Galileo with other positioning technologies (network-based, WLAN, etc.). Within these project, some pilot prototypes were implemented and demonstrated.

On the basis of the potential number of users, potential revenues for Galileo Operating Company or Concessionaire (GOC), international relevance, and level of innovation, a set of Priority Applications (PA) will be selected by the consortium and developed within the time-frame of the same project.

These applications will help to increase and optimise the use of the EGNOS services and the opportunities offered by the Galileo Signal Test-Bed (GSTB-V2) and the Galileo (IOV) phase.

Coins

The European Satellite Navigation project was selected as the main motif of a very high-value collectors' coin: the Austrian European Satellite Navigation commemorative coin, minted on 1 March 2006. The coin has a silver ring and gold-brown niobium "pill". In the reverse, the niobium portion depicts navigation satellites orbiting the Earth. The ring shows different modes of transport, for which satellite navigation was developed: an airplane, a car, a lorry, a train and a container ship.

Austrian €25 European Satellite Navigation commemorative coin, back

Receivers

A number of devices are compatible with Galileo. Samsung Galaxy S8 smartphones are compatible with Galileo, the first mainstream smartphones advertised with this capability.

Development Boards

The IOT4 Ltd created some development board for using the Galileo navigation network with Arduino, Raspberry or any Windows PC. They are available in different sizes and capabilities, based on Ublox M8 architecture. The GA-001 is LEA-M8, the GA-002 is MAX-M8 and the GA-003 is using NEO-M8 chipset.

Global Positioning System

The Global Positioning System (GPS), originally Navstar GPS, is a space-based radionavigation system owned by the United States government and operated by the United States Air Force. It is a global navigation satellite system that provides geolocation and time information to a GPS receiver anywhere on or near the Earth where there is an unobstructed line of sight to four or more GPS satellites.

Artist's conception of GPS Block II-F satellite in Earth orbit.

The GPS system does not require the user to transmit any data, and it operates independently of any telephonic or internet reception, though these technologies can enhance the usefulness of the GPS positioning information. The GPS system provides critical positioning capabilities to military, civil, and commercial users around the world. The United States government created the system, maintains it, and makes it freely accessible to anyone with a GPS receiver. However, the US government can selectively deny access to the system, as happened to the Indian military in 1999 during the Kargil War.

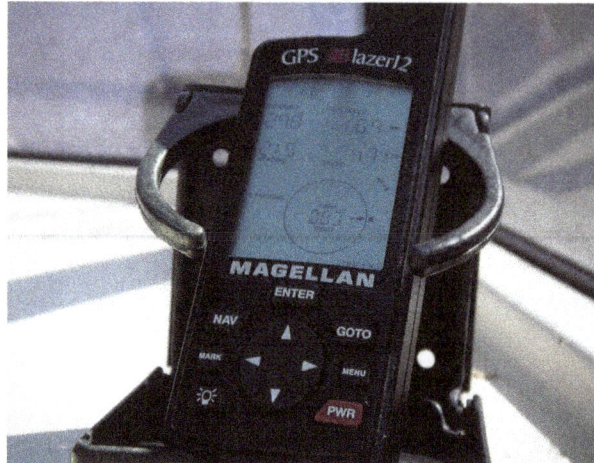

Civilian GPS receivers ("GPS navigation device") in a marine application.

The GPS project was launched in the United States in 1973 to overcome the limitations of previous navigation systems, integrating ideas from several predecessors, including a number of classified engineering design studies from the 1960s. The U.S. Department of Defense developed the system, which originally used 24 satellites. It became fully operational in 1995. Roger L. Easton of the Naval Research Laboratory, Ivan A. Getting of The Aerospace Corporation, and Bradford Parkinson of the Applied Physics Laboratory are credited with inventing it.

A U.S. Air Force Senior Airman runs through a checklist during Global Positioning System satellite operations.

Advances in technology and new demands on the existing system have now led to efforts to modernize the GPS and implement the next generation of GPS Block IIIA satellites and Next Generation Operational Control System (OCX). Announcements from Vice President Al Gore and the White House in 1998 initiated these changes. In 2000, the U.S. Congress authorized the modernization effort, GPS III.

In addition to GPS, other systems are in use or under development, mainly because of a potential denial of access by the US government. The Russian Global Navigation Satellite System (GLONASS) was developed contemporaneously with GPS, but suffered from incomplete coverage of the globe until the mid-2000s. GLONASS can be added to GPS devices, making more satellites available and enabling positions to be fixed more quickly and accurately, to within two meters. There are also the European Union Galileo positioning system, China's BeiDou Navigation Satellite System and India's NAVIC and Japan's Quasi-Zenith Satellite System.

History

The design of GPS is based partly on similar ground-based radio-navigation systems, such as LORAN and the Decca Navigator, developed in the early 1940s and used by the British Royal Navy during World War II.

Friedwardt Winterberg proposed a test of general relativity — detecting time slowing in a strong gravitational field using accurate atomic clocks placed in orbit inside artificial satellites.

Special and general relativity predict that the clocks on the GPS satellites would be seen by the Earth's observers to run 38 microseconds faster per day than the clocks on the Earth. The GPS calculated positions would quickly drift into error, accumulating to 10 kilometers per day. This was corrected for in the design of GPS.

Predecessors

The Soviet Union launched the first man-made satellite, Sputnik 1, in 1957. Two American physicists, William Guier and George Weiffenbach, at Johns Hopkins's Applied Physics Laboratory (APL), decided to monitor Sputnik's radio transmissions. Within hours they realized that, because of the Doppler effect, they could pinpoint where the satellite was along its orbit. The Director of the APL gave them access to their UNIVAC to do the heavy calculations required.

Official logo for NAVSTAR GPS

The next spring, Frank McClure, the deputy director of the APL, asked Guier and Weiffenbach to investigate the inverse problem — pinpointing the user's location, given that of the satellite. (At the time, the Navy was developing the submarine-launched Polaris missile, which required them to know the submarine's location.) This led them and APL to develop the TRANSIT system. In 1959, ARPA (renamed DARPA in 1972) also played a role in TRANSIT.

Emblem of the 50th Space Wing

The first satellite navigation system, TRANSIT, used by the United States Navy, was first successfully tested in 1960. It used a constellation of five satellites and could provide a navigational fix approximately once per hour.

In 1967, the U.S. Navy developed the Timation satellite, which proved the feasibility of placing accurate clocks in space, a technology required by GPS.

In the 1970s, the ground-based OMEGA navigation system, based on phase comparison of signal transmission from pairs of stations, became the first worldwide radio navigation system. Limitations of these systems drove the need for a more universal navigation solution with greater accuracy.

While there were wide needs for accurate navigation in military and civilian sectors, almost none of those was seen as justification for the billions of dollars it would cost in research, development, deployment, and operation for a constellation of navigation satellites. During the Cold War arms race, the nuclear threat to the existence of the United States was the one need that did justify this cost in the view of the United States Congress. This deterrent effect is why GPS was funded. It is also the reason for the ultra secrecy at that time. The nuclear triad consisted of the United States Navy's submarine-launched ballistic missiles (SLBMs) along with United States Air Force (USAF) strategic bombers and intercontinental ballistic missiles (ICBMs). Considered vital to the nuclear deterrence posture, accurate determination of the SLBM launch position was a force multiplier.

Precise navigation would enable United States ballistic missile submarines to get an accurate fix of their positions before they launched their SLBMs. The USAF, with two thirds of the nuclear triad, also had requirements for a more accurate and reliable navigation system. The Navy and Air Force were developing their own technologies in parallel to solve what was essentially the same problem.

To increase the survivability of ICBMs, there was a proposal to use mobile launch platforms (comparable to the Russian SS-24 and SS-25) and so the need to fix the launch position had similarity to the SLBM situation.

In 1960, the Air Force proposed a radio-navigation system called MOSAIC (MObile System for Accurate ICBM Control) that was essentially a 3-D LORAN. A follow-on study, Project 57, was worked in 1963 and it was "in this study that the GPS concept was born." That same year, the concept was pursued as Project 621B, which had "many of the attributes that you now see in GPS" and promised increased accuracy for Air Force bombers as well as ICBMs.

Updates from the Navy TRANSIT system were too slow for the high speeds of Air Force operation. The Naval Research Laboratory continued advancements with their Timation (Time Navigation) satellites, first launched in 1967, and with the third one in 1974 carrying the first atomic clock into orbit.

Another important predecessor to GPS came from a different branch of the United States military. In 1964, the United States Army orbited its first Sequential Collation of Range (SECOR) satellite used for geodetic surveying. The SECOR system included three ground-based transmitters from known locations that would send signals to the satellite transponder in orbit. A fourth ground-based station, at an undetermined position, could then use those signals to fix its location precisely. The last SECOR satellite was launched in 1969.

Decades later, during the early years of GPS, civilian surveying became one of the first fields to make use of the new technology, because surveyors could reap benefits of signals from the less-than-complete GPS constellation years before it was declared operational. GPS can be thought of as an evolution of the SECOR system where the ground-based transmitters have been migrated into orbit.

Development

With these parallel developments in the 1960s, it was realized that a superior system could be developed by synthesizing the best technologies from 621B, Transit, Timation, and SECOR in a multi-service program.

During Labor Day weekend in 1973, a meeting of about twelve military officers at the Pentagon discussed the creation of a *Defense Navigation Satellite System (DNSS)*. It was at this meeting that the real synthesis that became GPS was created. Later that year, the DNSS program was named *Navstar*, or Navigation System Using Timing and Ranging. With the individual satellites being associated with the name Navstar (as with the predecessors Transit and Timation), a more fully encompassing name was used to identify the constellation of Navstar satellites, *Navstar-GPS*. Ten "Block I" prototype satellites were launched between 1978 and 1985 (an additional unit was destroyed in a launch failure).

After Korean Air Lines Flight 007, a Boeing 747 carrying 269 people, was shot down in 1983 after straying into the USSR's prohibited airspace, in the vicinity of Sakhalin and Moneron Islands, President Ronald Reagan issued a directive making GPS freely available for civilian use, once it was sufficiently developed, as a common good. The first Block II satellite was launched on February 14, 1989, and the 24th satellite was launched in 1994. The GPS program cost at this point, not including the cost of the user equipment, but including the costs of the satellite launches, has been estimated at about USD 5 billion (then-year dollars). Roger L. Easton is widely credited as the primary inventor of GPS.

Initially, the highest quality signal was reserved for military use, and the signal available for civilian use was intentionally degraded (Selective Availability). This changed with President Bill Clinton signing a policy directive to turn off Selective Availability May 1, 2000 to provide the same accuracy to civilians that was afforded to the military. The directive was proposed by the U.S. Secretary of Defense, William Perry, because of the widespread growth of differential GPS services to improve civilian accuracy and eliminate the U.S. military advantage. Moreover, the U.S. military was actively developing technologies to deny GPS service to potential adversaries on a regional basis.

Since its deployment, the U.S. has implemented several improvements to the GPS service including new signals for civil use and increased accuracy and integrity for all users, all the while maintaining compatibility with existing GPS equipment. Modernization of the satellite system has been an ongoing initiative by the U.S. Department of Defense through a series of satellite acquisitions to meet the growing needs of the military, civilians, and the commercial market.

As of early 2015, high-quality, FAA grade, Standard Positioning Service (SPS) GPS receivers provide horizontal accuracy of better than 3.5 meters, although many factors such as receiver quality and atmospheric issues can affect this accuracy.

GPS is owned and operated by the United States government as a national resource. The Department of Defense is the steward of GPS. The *Interagency GPS Executive Board (IGEB)* oversaw GPS policy matters from 1996 to 2004. After that the National Space-Based Positioning, Navigation and Timing Executive Committee was established by presidential directive in 2004 to advise and coordinate federal departments and agencies on matters concerning the GPS and related systems. The executive committee is chaired jointly by the Deputy Secretaries of Defense and Transportation. Its membership includes equivalent-level officials from the Departments of State, Commerce, and Homeland Security, the Joint Chiefs of Staff and NASA. Components of the executive office of the president participate as observers to the executive committee, and the FCC chairman participates as a liaison.

The U.S. Department of Defense is required by law to "maintain a Standard Positioning Service (as defined in the federal radio navigation plan and the standard positioning service signal specification) that will be available on a continuous, worldwide basis," and "develop measures to prevent hostile use of GPS and its augmentations without unduly disrupting or degrading civilian uses."

Timeline and Modernization

Summary of Satellites						
Block	**Launch period**	**Satellite launches**				**Currently in orbit and healthy**
		Success	**Failure**	**In preparation**	**Planned**	
I	1978–1985	10	1	0	0	0
II	1989–1990	9	0	0	0	0
IIA	1990–1997	19	0	0	0	0
IIR	1997–2004	12	1	0	0	12
IIR-M	2005-2009	8	0	0	0	7
IIF	2010–2016	12	0	0	0	12
IIIA	From 2017	0	0	0	12	0
IIIB	—	0	0	0	8	0
IIIC	—	0	0	0	16	0
Total		70	2	0	36	31

- In 1972, the USAF Central Inertial Guidance Test Facility (Holloman AFB) conducted developmental flight tests of four prototype GPS receivers in a Y configuration over White Sands Missile Range, using ground-based pseudo-satellites.

- In 1978, the first experimental Block-I GPS satellite was launched.

- In 1983, after Soviet interceptor aircraft shot down the civilian airliner KAL 007 that strayed into prohibited airspace because of navigational errors, killing all 269 people on board, U.S. President Ronald Reagan announced that GPS would be made available for civilian uses once it was completed, although it had been previously published [in Navigation magazine] that the CA code (Coarse/Acquisition code) would be available to civilian users.

- By 1985, ten more experimental Block-I satellites had been launched to validate the concept.

- Beginning in 1988, Command & Control of these satellites was transitioned from Onizuka AFS, California to the 2nd Satellite Control Squadron (2SCS) located at Falcon Air Force Station in Colorado Springs, Colorado.

- On February 14, 1989, the first modern Block-II satellite was launched.

- The Gulf War from 1990 to 1991 was the first conflict in which the military widely used GPS.

- In 1991, a project to create a miniature GPS receiver successfully ended, replacing the previous 23 kg military receivers with a 1.25 kg handheld receiver.

- In 1992, the 2nd Space Wing, which originally managed the system, was inactivated and replaced by the 50th Space Wing.

- By December 1993, GPS achieved initial operational capability (IOC), indicating a full constellation (24 satellites) was available and providing the Standard Positioning Service (SPS).

- Full Operational Capability (FOC) was declared by Air Force Space Command (AFSPC) in April 1995, signifying full availability of the military's secure Precise Positioning Service (PPS).

- In 1996, recognizing the importance of GPS to civilian users as well as military users, U.S. President Bill Clinton issued a policy directive declaring GPS a dual-use system and establishing an Interagency GPS Executive Board to manage it as a national asset.

- In 1998, United States Vice President Al Gore announced plans to upgrade GPS with two new civilian signals for enhanced user accuracy and reliability, particularly with respect to aviation safety and in 2000 the United States Congress authorized the effort, referring to it as *GPS III*.

- On May 2, 2000 "Selective Availability" was discontinued as a result of the 1996 executive order, allowing users to receive a non-degraded signal globally.

- In 2004, the United States Government signed an agreement with the European Community establishing cooperation related to GPS and Europe's Galileo system.

- In 2004, United States President George W. Bush updated the national policy and replaced the executive board with the National Executive Committee for Space-Based Positioning, Navigation, and Timing.

- November 2004, Qualcomm announced successful tests of assisted GPS for mobile phones.

- In 2005, the first modernized GPS satellite was launched and began transmitting a second civilian signal (L2C) for enhanced user performance.

- On September 14, 2007, the aging mainframe-based Ground Segment Control System was transferred to the new Architecture Evolution Plan.

- On May 19, 2009, the United States Government Accountability Office issued a report warning that some GPS satellites could fail as soon as 2010.

- On May 21, 2009, the Air Force Space Command allayed fears of GPS failure, saying "There's only a small risk we will not continue to exceed our performance standard."

- On January 11, 2010, an update of ground control systems caused a software incompatibility with 8000 to 10000 military receivers manufactured by a division of Trimble Navigation Limited of Sunnyvale, Calif.

- On February 25, 2010, the U.S. Air Force awarded the contract to develop the GPS Next Generation Operational Control System (OCX) to improve accuracy and availability of GPS navigation signals, and serve as a critical part of GPS modernization.

Awards

On February 10, 1993, the National Aeronautic Association selected the GPS Team as winners of the 1992 Robert J. Collier Trophy, the nation's most prestigious aviation award. This team combines researchers from the Naval Research Laboratory, the USAF, the Aerospace Corporation, Rockwell International Corporation, and IBM Federal Systems Company. The citation honors them "for the most significant development for safe and efficient navigation and surveillance of air and spacecraft since the introduction of radio navigation 50 years ago."

Two GPS developers received the National Academy of Engineering Charles Stark Draper Prize for 2003:

- Ivan Getting, emeritus president of The Aerospace Corporation and an engineer at the Massachusetts Institute of Technology, established the basis for GPS, improving on the World War II land-based radio system called LORAN (*L*ong-range *R*adio *A*id to *N*avigation).

- Bradford Parkinson, professor of aeronautics and astronautics at Stanford University, conceived the present satellite-based system in the early 1960s and developed it in conjunction with the U.S. Air Force. Parkinson served twenty-one years in the Air Force, from 1957 to 1978, and retired with the rank of colonel.

GPS developer Roger L. Easton received the National Medal of Technology on February 13, 2006.

Francis X. Kane (Col. USAF, ret.) was inducted into the U.S. Air Force Space and Missile Pioneers Hall of Fame at Lackland A.F.B., San Antonio, Texas, March 2, 2010 for his role in space technology development and the engineering design concept of GPS conducted as part of Project 621B.

In 1998, GPS technology was inducted into the Space Foundation Space Technology Hall of Fame.

On October 4, 2011, the International Astronautical Federation (IAF) awarded the Global Positioning System (GPS) its 60th Anniversary Award, nominated by IAF member, the American Institute for Aeronautics and Astronautics (AIAA). The IAF Honors and Awards Committee recognized the uniqueness of the GPS program and the exemplary role it has played in building international collaboration for the benefit of humanity.

Basic Concept of GPS

Fundamentals

The GPS concept is based on time and the known position of specialized satellites. The satellites carry very stable atomic clocks that are synchronized with one another and to ground clocks. Any drift from true time maintained on the ground is corrected daily. Likewise, the satellite locations are known with great precision. GPS receivers have clocks as well; however, they are usually not synchronized with true time, and are less stable. GPS satellites continuously transmit their current time and position. A GPS receiver monitors multiple satellites and solves equations to determine the precise position of the receiver and its deviation from true time. At a minimum, four satellites must be in view of the receiver for it to compute four unknown quantities (three position coordinates and clock deviation from satellite time).

More Detailed Description

Each GPS satellite continually broadcasts a signal (carrier wave with modulation) that includes:

- A pseudorandom code (sequence of ones and zeros) that is known to the receiver. By time-aligning a receiver-generated version and the receiver-measured version of the code, the time of arrival (TOA) of a defined point in the code sequence, called an epoch, can be found in the receiver clock time scale

- A message that includes the time of transmission (TOT) of the code epoch (in GPS system time scale) and the satellite position at that time

Conceptually, the receiver measures the TOAs (according to its own clock) of four satellite signals. From the TOAs and the TOTs, the receiver forms four time of flight (TOF) values, which are (given the speed of light) approximately equivalent to receiver-satellite range differences. The receiver then computes its three-dimensional position and clock deviation from the four TOFs.

In practice the receiver position (in three dimensional Cartesian coordinates with origin at the Earth's center) and the offset of the receiver clock relative to the GPS time are computed simultaneously, using the navigation equations to process the TOFs.

The receiver's Earth-centered solution location is usually converted to latitude, longitude and height relative to an ellipsoidal Earth model. The height may then be further converted to height relative to the geoid (e.g., EGM96) (essentially, mean sea level). These coordinates may be displayed, e.g., on a moving map display, and/or recorded and/or used by some other system (e.g., a vehicle guidance system).

User-satellite Geometry

Although usually not formed explicitly in the receiver processing, the conceptual time differences of arrival (TDOAs) define the measurement geometry. Each TDOA corresponds to a hyperboloid of revolution. The line connecting the two satellites involved (and its extensions) forms the axis of the hyperboloid. The receiver is located at the point where three hyperboloids intersect.

It is sometimes incorrectly said that the user location is at the intersection of three spheres. While simpler to visualize, this is only the case if the receiver has a clock synchronized with the satellite clocks (i.e., the receiver measures true ranges to the satellites rather than range differences). There are significant performance benefits to the user carrying a clock synchronized with the satellites. Foremost is that only three satellites are needed to compute a position solution. If this were part of the GPS system concept so that all users needed to carry a synchronized clock, then a smaller number of satellites could be deployed. However, the cost and complexity of the user equipment would increase significantly.

Receiver in Continuous Operation

The description above is representative of a receiver start-up situation. Most receivers have a track algorithm, sometimes called a *tracker*, that combines sets of satellite measurements collected at different times—in effect, taking advantage of the fact that successive receiver positions are usually close to each other. After a set of measurements are processed, the tracker predicts the receiver location corresponding to the next set of satellite measurements. When the new measurements are collected, the receiver uses a weighting scheme to combine the new measurements with the tracker prediction. In general, a tracker can (a) improve receiver position and time accuracy, (b) reject bad measurements, and (c) estimate receiver speed and direction.

The disadvantage of a tracker is that changes in speed or direction can only be computed with a delay, and that derived direction becomes inaccurate when the distance traveled between two position measurements drops below or near the random error of position measurement. GPS units can use measurements of the Doppler shift of the signals received to compute velocity accurately. More advanced navigation systems use additional sensors like a compass or an inertial navigation system to complement GPS.

Non-navigation Applications

In typical GPS operation as a navigator, four or more satellites must be visible to obtain an accurate result. The solution of the navigation equations gives the position of the receiver along with the difference between the time kept by the receiver's on-board clock and the true time-of-day, thereby eliminating the need for a more precise and possibly impractical receiver based clock. Applications for GPS such as time transfer, traffic signal timing, and synchronization of cell phone base stations, make use of this cheap and highly accurate timing. Some GPS applications use this time for display, or, other than for the basic position calculations, do not use it at all.

Although four satellites are required for normal operation, fewer apply in special cases. If one variable is already known, a receiver can determine its position using only three satellites. For example, a ship or aircraft may have known elevation. Some GPS receivers may use additional clues or assumptions such as reusing the last known altitude, dead reckoning, inertial navigation, or including information from the vehicle computer, to give a (possibly degraded) position when fewer than four satellites are visible.

Structure

The current GPS consists of three major segments. These are the space segment (SS), a control segment (CS), and a user segment (US). The U.S. Air Force develops, maintains, and operates the space and control segments. GPS satellites broadcast signals from space, and each GPS receiver uses these signals to calculate its three-dimensional location (latitude, longitude, and altitude) and the current time.

The space segment is composed of 24 to 32 satellites in medium Earth orbit and also includes the payload adapters to the boosters required to launch them into orbit. The control segment is composed of a master control station (MCS), an alternate master control station, and a host of dedicated and shared ground antennas and monitor stations. The user segment is composed of hundreds of thousands of U.S. and allied military users of the secure GPS Precise Positioning Service, and hundreds of millions of civil, commercial, and scientific users of the Standard Positioning Service.

Space Segment

Unlaunched GPS block II-A satellite on display at the San Diego Air & Space Museum

The space segment (SS) is composed of the orbiting GPS satellites, or Space Vehicles (SV) in GPS parlance. The GPS design originally called for 24 SVs, eight each in three approximately circular orbits, but this was modified to six orbital planes with four satellites each. The six orbit planes have approximately 55° inclination (tilt relative to the Earth's equator) and are separated by 60° right ascension of the ascending node (angle along the equator from a reference point to the orbit's intersection). The orbital period is one-half a sidereal day, i.e., 11 hours and 58 minutes so that the satellites pass over the same locations or almost the same locations every day. The orbits are arranged so that at least six satellites are always within line of sight from almost everywhere on the Earth's surface. The result of this objective is that the four satellites are not evenly spaced (90 degrees) apart within each orbit. In general terms, the angular difference between satellites in each orbit is 30, 105, 120, and 105 degrees apart, which sum to 360 degrees.

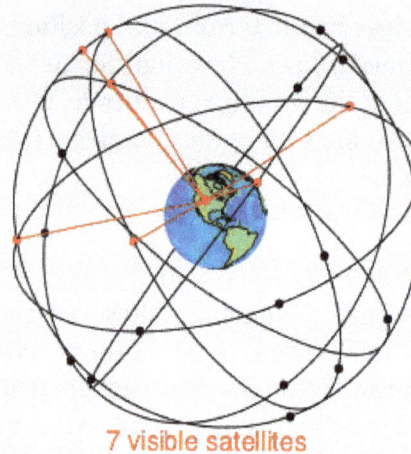

7 visible satellites

A visual example of a 24 satellite GPS constellation in motion with the earth rotating. Notice how the number of *satellites in view* from a given point on the earth's surface, in this example in Golden, Colorado, USA(39.7469° N, 105.2108° W), changes with time.

Orbiting at an altitude of approximately 20,200 km (12,600 mi); orbital radius of approximately 26,600 km (16,500 mi), each SV makes two complete orbits each sidereal day, repeating the same ground track each day. This was very helpful during development because even with only four satellites, correct alignment means all four are visible from one spot for a few hours each day. For military operations, the ground track repeat can be used to ensure good coverage in combat zones.

As of February 2016, there are 32 satellites in the GPS constellation, 31 of which are in use. The additional satellites improve the precision of GPS receiver calculations by providing redundant measurements. With the increased number of satellites, the constellation was changed to a non-uniform arrangement. Such an arrangement was shown to improve reliability and availability of the system, relative to a uniform system, when multiple satellites fail. About nine satellites are visible from any point on the ground at any one time. ensuring considerable redundancy over the minimum four satellites needed for a position.

Control Segment

Ground monitor station used from 1984 to 2007, on display at the Air Force Space & Missile Museum.

The control segment is composed of:

1. a master control station (MCS),

2. an alternate master control station,

3. four dedicated ground antennas, and

4. six dedicated monitor stations.

The MCS can also access U.S. Air Force Satellite Control Network (AFSCN) ground antennas (for additional command and control capability) and NGA (National Geospatial-Intelligence Agency) monitor stations. The flight paths of the satellites are tracked by dedicated U.S. Air Force monitoring stations in Hawaii, Kwajalein Atoll, Ascension Island, Diego Garcia, Colorado Springs, Colorado and Cape Canaveral, along with shared NGA monitor stations operated in England, Argentina, Ecuador, Bahrain, Australia and Washington DC. The tracking information is sent to the Air Force Space Command MCS at Schriever Air Force Base 25 km (16 mi) ESE of Colorado Springs, which is operated by the 2nd Space Operations Squadron (2 SOPS) of the U.S. Air Force. Then 2 SOPS contacts each GPS satellite regularly with a navigational update using dedicated or shared (AFSCN) ground antennas (GPS dedicated ground antennas are located at Kwajalein, Ascension Island, Diego Garcia, and Cape Canaveral). These updates synchronize the atomic clocks on board the satellites to within a few nanoseconds of each other, and adjust the ephemeris of each satellite's internal orbital model. The updates are created by a Kalman filter that uses inputs from the ground monitoring stations, space weather information, and various other inputs.

Satellite maneuvers are not precise by GPS standards—so to change a satellite's orbit, the satellite must be marked *unhealthy*, so receivers don't use it. After the satellite maneuver, engineers track the new orbit from the ground, upload the new ephemeris, and mark the satellite healthy again.

The Operation Control Segment (OCS) currently serves as the control segment of record. It provides the operational capability that supports GPS users and keeps the GPS system operational and performing within specification.

OCS successfully replaced the legacy 1970s-era mainframe computer at Schriever Air Force Base in September 2007. After installation, the system helped enable upgrades and provide a foundation for a new security architecture that supported U.S. armed forces. OCS will continue to be the ground control system of record until the new segment, Next Generation GPS Operation Control System (OCX), is fully developed and functional.

The new capabilities provided by OCX will be the cornerstone for revolutionizing GPS's mission capabilities, enabling Air Force Space Command to greatly enhance GPS operational services to U.S. combat forces, civil partners and myriad domestic and international users.

The GPS OCX program also will reduce cost, schedule and technical risk. It is designed to provide 50% sustainment cost savings through efficient software architecture and Performance-Based Logistics. In addition, GPS OCX is expected to cost millions less than the cost to upgrade OCS while providing four times the capability.

The GPS OCX program represents a critical part of GPS modernization and provides significant information assurance improvements over the current GPS OCS program.

- OCX will have the ability to control and manage GPS legacy satellites as well as the next generation of GPS III satellites, while enabling the full array of military signals.

- Built on a flexible architecture that can rapidly adapt to the changing needs of today's and future GPS users allowing immediate access to GPS data and constellation status through secure, accurate and reliable information.

- Provides the warfighter with more secure, actionable and predictive information to enhance situational awareness.

- Enables new modernized signals (L1C, L2C, and L5) and has M-code capability, which the legacy system is unable to do.

- Provides significant information assurance improvements over the current program including detecting and preventing cyber attacks, while isolating, containing and operating during such attacks.

- Supports higher volume near real-time command and control capabilities and abilities.

On September 14, 2011, the U.S. Air Force announced the completion of GPS OCX Preliminary Design Review and confirmed that the OCX program is ready for the next phase of development.

The GPS OCX program has missed major milestones and is pushing the GPS IIIA launch beyond April 2016.

User Segment

GPS receivers come in a variety of formats, from devices integrated into cars, phones, and watches, to dedicated devices such as these.

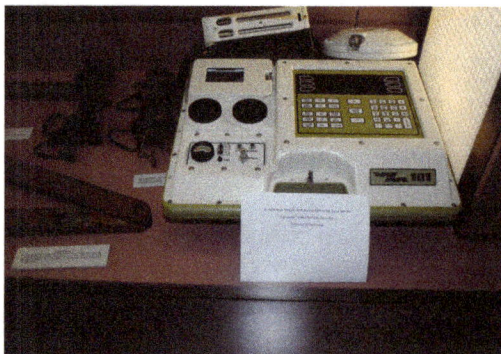

The first portable GPS unit, Leica WM 101 displayed at the Irish National Science Museum at Maynooth.

The user segment is composed of hundreds of thousands of U.S. and allied military users of the secure GPS Precise Positioning Service, and tens of millions of civil, commercial and scientific users of the Standard Positioning Service. In general, GPS receivers are composed of an antenna, tuned to the frequencies transmitted by the satellites, receiver-processors, and a highly stable clock (often a crystal oscillator). They may also include a display for providing location and speed information to the user. A receiver is often described by its number of channels: this signifies how many satellites it can monitor simultaneously. Originally limited to four or five, this has progressively increased over the years so that, as of 2007, receivers typically have between 12 and 20 channels. Though there are many receiver manufacturers, they almost all use one of the chipsets produced for this purpose.

A typical OEM GPS receiver module measuring 15×17 mm.

GPS receivers may include an input for differential corrections, using the RTCM SC-104 format. This is typically in the form of an RS-232 port at 4,800 bit/s speed. Data is actually sent at a much lower rate, which limits the accuracy of the signal sent using RTCM. Receivers with internal DGPS receivers can outperform those using external RTCM data. As of 2006, even low-cost units commonly include Wide Area Augmentation System (WAAS) receivers.

Many GPS receivers can relay position data to a PC or other device using the NMEA 0183 protocol. Although this protocol is officially defined by the National Marine Electronics Association (NMEA), references to this protocol have been compiled from public records, allowing open source tools like gpsd to read the protocol without violating intellectual property laws. Other proprietary protocols exist as well, such as the SiRF and MTK protocols. Receivers can interface with other devices using methods including a serial connection, USB, or Bluetooth.

Applications

While originally a military project, GPS is considered a *dual-use* technology, meaning it has significant military and civilian applications.

GPS has become a widely deployed and useful tool for commerce, scientific uses, tracking, and surveillance. GPS's accurate time facilitates everyday activities such as banking, mobile phone operations, and even the control of power grids by allowing well synchronized hand-off switching.

Civilian

Many civilian applications use one or more of GPS's three basic components: absolute location, relative movement, and time transfer.

This antenna is mounted on the roof of a hut containing a scientific experiment needing precise timing.

- Agriculture: GPS has made a great evolution in different aspects of modern agricultural sectors. Today, a growing number of crop producers are using GPS and other modern electronic and computer equipment to practice Site Specific Management (SSM) and precision agriculture. This technology has the potential in agricultural mechanization (farm and machinery management) by providing farmers with a sophisticated tool to measure yield on much smaller scales as well as precise determination and automatic storing of variables such as field time, working area, machine travel distance and speed, fuel consumption and yield information.

- Astronomy: both positional and clock synchronization data is used in astrometry and celestial mechanics. GPS is also used in both amateur astronomy with small telescopes as well as by professional observatories for finding extrasolar planets, for example.

- Automated vehicle: applying location and routes for cars and trucks to function without a human driver.

- Cartography: both civilian and military cartographers use GPS extensively.

- Cellular telephony: clock synchronization enables time transfer, which is critical for synchronizing its spreading codes with other base stations to facilitate inter-cell handoff and support hybrid GPS/cellular position detection for mobile emergency calls and other applications. The first handsets with integrated GPS launched in the late 1990s. The U.S. Federal Communications Commission (FCC) mandated the feature in either the handset or in the towers (for use in triangulation) in 2002 so emergency services could locate 911 callers. Third-party software developers later gained access to GPS APIs from Nextel upon launch, followed by Sprint in 2006, and Verizon soon thereafter.

- Clock synchronization: the accuracy of GPS time signals (±10 ns) is second only to the atomic clocks they are based on.

- Disaster relief/emergency services: many emergency services depend upon GPS for location and timing capabilities.

- GPS-equipped radiosondes and dropsondes: measure and calculate the atmospheric pressure, wind speed and direction up to 27 km from the Earth's surface.

- Radio occultation for weather and atmospheric science applications.

- Fleet tracking: used to identify, locate and maintain contact reports with one or more fleet vehicles in real-time.

- Geofencing: vehicle tracking systems, person tracking systems, and pet tracking systems use GPS to locate devices that are attached to or carried by a person, vehicle, or pet. The application can provide continuous tracking and send notifications if the target leaves a designated (or "fenced-in") area.

- Geotagging: applies location coordinates to digital objects such as photographs (in Exif data) and other documents for purposes such as creating map overlays with devices like Nikon GP-1

- GPS aircraft tracking

- GPS for mining: the use of RTK GPS has significantly improved several mining operations such as drilling, shoveling, vehicle tracking, and surveying. RTK GPS provides centimeter-level positioning accuracy.

- GPS data mining: It is possible to aggregate GPS data from multiple users to understand movement patterns, common trajectories and interesting locations.

- GPS tours: location determines what content to display; for instance, information about an approaching point of interest.

- Navigation: navigators value digitally precise velocity and orientation measurements.

- Phasor measurements: GPS enables highly accurate timestamping of power system measurements, making it possible to compute phasors.

- Recreation: for example, Geocaching, Geodashing, GPS drawing, waymarking, and other kinds of location based mobile games.

- Robotics: self-navigating, autonomous robots using a GPS sensors, which calculate latitude, longitude, time, speed, and heading.

- Sport: used in football and rugby for the control and analysis of the training load.

- Surveying: surveyors use absolute locations to make maps and determine property boundaries.

- Tectonics: GPS enables direct fault motion measurement of earthquakes. Between earthquakes GPS can be used to measure crustal motion and deformation to estimate seismic strain buildup for creating seismic hazard maps.

- Telematics: GPS technology integrated with computers and mobile communications technology in automotive navigation systems.

Restrictions on Civilian Use

The U.S. government controls the export of some civilian receivers. All GPS receivers capable of functioning above 18 km (60,000 feet) altitude and 515 m/s (1,000 knots), or designed or modified for use with unmanned air vehicles like, e.g., ballistic or cruise missile systems, are classified as munitions (weapons)—which means they require State Department export licenses.

This rule applies even to otherwise purely civilian units that only receive the L1 frequency and the C/A (Coarse/Acquisition) code.

Disabling operation above these limits exempts the receiver from classification as a munition. Vendor interpretations differ. The rule refers to operation at both the target altitude and speed, but some receivers stop operating even when stationary. This has caused problems with some amateur radio balloon launches that regularly reach 30 km (100,000 feet).

These limits only apply to units or components exported from the USA. A growing trade in various components exists, including GPS units from other countries. These are expressly sold as ITAR-free.

Military

M982 Excalibur GPS-guided artillery shell.

As of 2009, military GPS applications include:

- Navigation: Soldiers use GPS to find objectives, even in the dark or in unfamiliar territory, and to coordinate troop and supply movement. In the United States armed forces, commanders use the *Commander's Digital Assistant* and lower ranks use the *Soldier Digital Assistant*.

- Target tracking: Various military weapons systems use GPS to track potential ground and air targets before flagging them as hostile. These weapon systems pass target coordinates to precision-guided munitions to allow them to engage targets accurately. Military aircraft, particularly in air-to-ground roles, use GPS to find targets.

- Missile and projectile guidance: GPS allows accurate targeting of various military weapons including ICBMs, cruise missiles, precision-guided munitions and artillery shells. Embedded GPS receivers able to withstand accelerations of 12,000 g or about 118 km/s² have been developed for use in 155-millimeter (6.1 in) howitzer shells.

- Search and rescue.

- Reconnaissance: Patrol movement can be managed more closely.

- GPS satellites carry a set of nuclear detonation detectors consisting of an optical sensor (Y-sensor), an X-ray sensor, a dosimeter, and an electromagnetic pulse (EMP) sensor (W-sensor), that form a major portion of the United States Nuclear Detonation Detection System. General William Shelton has stated that future satellites may drop this feature to save money.

GPS type navigation was first used in war in the 1991 Persian Gulf War, before GPS was fully developed in 1995, to assist Coalition Forces to navigate and perform maneuvers in the war. The war also demonstrated the vulnerability of GPS to being jammed, when Iraqi forces added noise to the weak GPS signal transmission to protect Iraqi targets.

Communication

The navigational signals transmitted by GPS satellites encode a variety of information including satellite positions, the state of the internal clocks, and the health of the network. These signals are transmitted on two separate carrier frequencies that are common to all satellites in the network. Two different encodings are used: a public encoding that enables lower resolution navigation, and an encrypted encoding used by the U.S. military.

Message Format

Each GPS satellite continuously broadcasts a *navigation message* on L1 (C/A and P/Y) and L2 (P/Y) frequencies at a rate of 50 bits per second. Each complete message takes 750 seconds (12 1/2 minutes) to complete. The message structure has a basic format of a 1500-bit-long frame made up of five subframes, each subframe being 300 bits (6 seconds) long. Subframes 4 and 5 are subcommutated 25 times each, so that a complete data message requires the transmission of 25 full frames. Each subframe consists of ten words, each 30 bits long. Thus, with 300 bits in a subframe times 5 subframes in a frame times 25 frames in a message, each message is 37,500 bits long. At a transmission rate of 50-bit/s, this gives 750 seconds to transmit an entire almanac message (GPS). Each 30-second frame begins precisely on the minute or half-minute as indicated by the atomic clock on each satellite.

The first subframe of each frame encodes the week number and the time within the week, as well as the data about the health of the satellite. The second and the third subframes contain the *ephemeris* – the precise orbit for the satellite. The fourth and fifth subframes contain the *almanac*, which contains coarse orbit and status information for up to 32 satellites in the constellation as well as data related to error correction. Thus, to obtain an accurate satellite location from this transmitted message, the receiver must demodulate the message from each satellite it includes in its solution for 18 to 30 seconds. To collect all transmitted almanacs, the receiver must demodulate the message for 732 to 750 seconds or 12 1/2 minutes.

All satellites broadcast at the same frequencies, encoding signals using unique code division multiple access (CDMA) so receivers can distinguish individual satellites from each other. The system uses two distinct CDMA encoding types: the coarse/acquisition (C/A) code, which is accessible by the general public, and the precise (P(Y)) code, which is encrypted so that only the U.S. military and other NATO nations who have been given access to the encryption code can access it.

The ephemeris is updated every 2 hours and is generally valid for 4 hours, with provisions for updates every 6 hours or longer in non-nominal conditions. The almanac is updated typically every 24 hours. Additionally, data for a few weeks following is uploaded in case of transmission updates that delay data upload.

Satellite Frequencies

All satellites broadcast at the same two frequencies, 1.57542 GHz (L1 signal) and 1.2276 GHz (L2 signal). The satellite network uses a CDMA spread-spectrum technique where the low-bitrate message data is encoded with a high-rate pseudo-random (PRN) sequence that is different for each satellite. The receiver must be aware of the PRN codes for each satellite to reconstruct the actual message data. The C/A code, for civilian use, transmits data at 1.023 million chips per second, whereas the P code, for U.S. military use, transmits at 10.23 million chips per second. The actual internal reference of the satellites is 10.22999999543 MHz to compensate for relativistic effects that make observers on the Earth perceive a different time reference with respect to the transmitters in orbit. The L1 carrier is modulated by both the C/A and P codes, while the L2 carrier is only modulated by the P code. The P code can be encrypted as a so-called P(Y) code that is only available to military equipment with a proper decryption key. Both the C/A and P(Y) codes impart the precise time-of-day to the user.

The L3 signal at a frequency of 1.38105 GHz is used to transmit data from the satellites to ground stations. This data is used by the United States Nuclear Detonation (NUDET) Detection System (USNDS) to detect, locate, and report nuclear detonations (NUDETs) in the Earth's atmosphere and near space. One usage is the enforcement of nuclear test ban treaties.

The L4 band at 1.379913 GHz is being studied for additional ionospheric correction.

The L5 frequency band at 1.17645 GHz was added in the process of GPS modernization. This frequency falls into an internationally protected range for aeronautical navigation, promising little or no interference under all circumstances. The first Block IIF satellite that provides this signal was launched in 2010. The L5 consists of two carrier components that are in phase quadrature with each other. Each carrier component is bi-phase shift key (BPSK) modulated by a separate bit train. "L5, the third civil GPS signal, will eventually support safety-of-life applications for aviation and provide improved availability and accuracy."

A conditional waiver has recently (2011-01-26) been granted to LightSquared to operate a terrestrial broadband service near the L1 band. Although LightSquared had applied for a license to operate in the 1525 to 1559 band as early as 2003 and it was put out for public comment, the FCC asked LightSquared to form a study group with the GPS community to test GPS receivers and identify issue that might arise due to the larger signal power from the LightSquared terrestrial network. The

GPS community had not objected to the LightSquared (formerly MSV and SkyTerra) applications until November 2010, when LightSquared applied for a modification to its Ancillary Terrestrial Component (ATC) authorization. This filing (SAT-MOD-20101118-00239) amounted to a request to run several orders of magnitude more power in the same frequency band for terrestrial base stations, essentially repurposing what was supposed to be a "quiet neighborhood" for signals from space as the equivalent of a cellular network. Testing in the first half of 2011 has demonstrated that the impact of the lower 10 MHz of spectrum is minimal to GPS devices (less than 1% of the total GPS devices are affected). The upper 10 MHz intended for use by LightSquared may have some impact on GPS devices. There is some concern that this may seriously degrade the GPS signal for many consumer uses. Aviation Week magazine reports that the latest testing (June 2011) confirms "significant jamming" of GPS by LightSquared's system.

Demodulation and Decoding

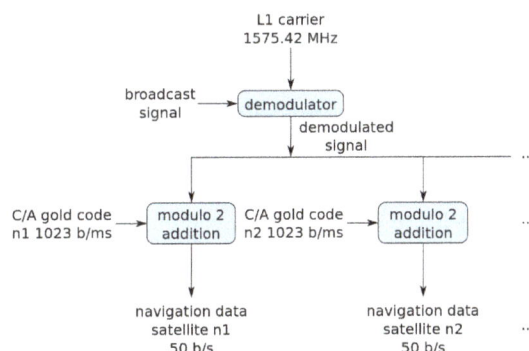

Demodulating and Decoding GPS Satellite Signals using the Coarse/Acquisition Gold code.

Because all of the satellite signals are modulated onto the same L1 carrier frequency, the signals must be separated after demodulation. This is done by assigning each satellite a unique binary sequence known as a Gold code. The signals are decoded after demodulation using addition of the Gold codes corresponding to the satellites monitored by the receiver.

If the almanac information has previously been acquired, the receiver picks the satellites to listen for by their PRNs, unique numbers in the range 1 through 32. If the almanac information is not in memory, the receiver enters a search mode until a lock is obtained on one of the satellites. To obtain a lock, it is necessary that there be an unobstructed line of sight from the receiver to the satellite. The receiver can then acquire the almanac and determine the satellites it should listen for. As it detects each satellite's signal, it identifies it by its distinct C/A code pattern. There can be a delay of up to 30 seconds before the first estimate of position because of the need to read the ephemeris data.

Processing of the navigation message enables the determination of the time of transmission and the satellite position at this time.

Problem Description

The receiver uses messages received from satellites to determine the satellite positions and time sent. The x, y, and z components of satellite position and the time sent are designated as $[x_i, y_i, z_i, s_i]$ where the subscript i denotes the satellite and has the value 1, 2, ..., n, where $n \geq 4$. When the time of message reception indicated by the on-board receiver clock is $\tilde{t}_i$, the true reception time is $t_i = \tilde{t}_i - b$, where b is the receiver's clock bias from the much more accurate GPS system clocks employed by the satellites. The receiver clock bias is the same for all received satellite signals (assuming the satellite clocks are all perfectly synchronized). The message's transit time is $\tilde{t}_i - b - s_i$, where s_i is the satellite time. Assuming the message traveled at the speed of light, c, the distance traveled is $(\tilde{t}_i - b - s_i)\,c$.

For n satellites, the equations to satisfy are:

$$(x - x_i)^2 + (y - y_i)^2 + (z - z_i)^2 = \left([\tilde{t}_i - b - s_i]c\right)^2, i = 1, 2, \ldots, n$$

or in terms of *pseudoranges*, $p_i = \left(\tilde{t}_i - s_i\right)c,$ as

$$\sqrt{(x - x_i)^2 + (y - y_i)^2 + (z - z_i)^2} + bc = p_i, i = 1, 2, \ldots, n.$$

Since the equations have four unknowns [x, y, z, b]—the three components of GPS receiver position and the clock bias—signals from at least four satellites are necessary to attempt solving these equations. They can be solved by algebraic or numerical methods. Existence and uniqueness of GPS solutions are discussed by Abell and Chaffee. When n is greater than 4 this system is overdetermined and a fitting method must be used.

With each combination of satellites, GDOP quantities can be calculated based on the relative sky directions of the satellites used. The receiver location is expressed in a specific coordinate system, such as latitude and longitude using the WGS 84 geodetic datum or a country-specific system.

Geometric Interpretation

The GPS equations can be solved by numerical and analytical methods. Geometrical interpretations can enhance the understanding of these solution methods.

Spheres

The measured ranges, called pseudoranges, contain clock errors. In a simplified idealization in which the ranges are synchronized, these true ranges represent the radii of spheres, each centered on one of the transmitting satellites. The solution for the position of the receiver is then at the intersection of the surfaces of three of these spheres. If more than the minimum number of ranges is available, a near intersection of more than three sphere surfaces could be found via, e.g. least squares.

Hyperboloids

If the distance traveled between the receiver and satellite i and the distance traveled between the receiver and satellite j are subtracted, the result is $(\tilde{t}_i - s_i)\,c - (\tilde{t}_j - s_j)\,c$, which only involves known or measured quantities. The locus of points having a constant difference in distance to two points (here, two satellites) is a hyperboloid. Thus, from four or more measured reception times, the receiver can be placed at the intersection of the surfaces of three or more hyperboloids.

Spherical Cones

The solution space [x, y, z, b] can be seen as a four-dimensional geometric space. In that case each of the equations describes a spherical cone, with the cusp located at the satellite, and the base a sphere around the satellite. The receiver is at the intersection of four or more of such cones.

Solution Methods

Least Squares

When more than four satellites are available, the calculation can use the four best, or more than four simultaneously (up to all visible satellites), depending on the number of receiver channels, processing capability, and geometric dilution of precision (GDOP).

Using more than four involves an over-determined system of equations with no unique solution; such a system can be solved by a least-squares or weighted least squares method.

$$\left(\hat{x},\hat{y},\hat{z},\hat{b}\right)=\underset{(x,y,z,b)}{\arg\min}\sum_i\left(\sqrt{(x-x_i)^2+(y-y_i)^2+(z-z_i)^2}+bc-p_i\right)^2$$

Iterative

Both the equations for four satellites, or the least squares equations for more than four, are non-linear and need special solution methods. A common approach is by iteration on a linearized form of the equations, such as the Gauss–Newton algorithm.

The GPS system was initially developed assuming use of a numerical least-squares solution method—i.e., before closed-form solutions were found.

Closed-form

One closed-form solution to the above set of equations was developed by S. Bancroft. Its properties are well known; in particular, proponents claim it is superior in low-GDOP situations, compared to iterative least squares methods.

Bancroft's method is algebraic, as opposed to numerical, and can be used for four or more satellites. When four satellites are used, the key steps are inversion of a 4x4 matrix and solution of a single-variable quadratic equation. Bancroft's method provides one or two solutions for the unknown quantities. When there are two (usually the case), only one is a near-Earth sensible solution.

When a receiver uses more than four satellites for a solution, Bancroft uses the generalized inverse (i.e., the pseudoinverse) to find a solution. However, a case has been made that iterative methods (e.g., Gauss–Newton algorithm) for solving over-determined non-linear least squares (NLLS) problems generally provide more accurate solutions.

Leick et al. (2015) states that "Bancroft's (1985) solution is a very early, if not the first, closed-form solution." Other closed-form solutions were published afterwards, although their adoption in practice is unclear.

Error Sources and Analysis

GPS error analysis examines error sources in GPS results and the expected size of those errors. GPS makes corrections for receiver clock errors and other effects, but some residual errors remain uncorrected. Error sources include signal arrival time measurements, numerical calculations, atmospheric effects (ionospheric/tropospheric delays), ephemeris and clock data, multipath signals, and natural and artificial interference. Magnitude of residual errors from these sources depends on geometric dilution of precision. Artificial errors may result from jamming devices and threaten ships and aircraft or from intentional signal degradation through selective availability, which limited accuracy to ≈6–12 m, but has been switched off since May 1, 2000.

Augmentation

Integrating external information into the calculation process can materially improve accuracy. Such augmentation systems are generally named or described based on how the information arrives. Some systems transmit additional error information (such as clock drift, ephemera, or ionospheric delay), others characterize prior errors, while a third group provides additional navigational or vehicle information.

Examples of augmentation systems include the Wide Area Augmentation System (WAAS), European Geostationary Navigation Overlay Service (EGNOS), Differential GPS (DGPS), inertial navigation systems (INS) and Assisted GPS. The standard accuracy of about 15 meters (49 feet) can be augmented to 3–5 meters (9.8–16.4 ft) with DGPS, and to about 3 meters (9.8 feet) with WAAS.

Precise Monitoring

Accuracy can be improved through precise monitoring and measurement of existing GPS signals in additional or alternate ways.

The largest remaining error is usually the unpredictable delay through the ionosphere. The spacecraft broadcast ionospheric model parameters, but some errors remain. This is one reason GPS spacecraft transmit on at least two frequencies, L1 and L2. Ionospheric delay is a well-defined function of frequency and the total electron content (TEC) along the path, so measuring the arrival time difference between the frequencies determines TEC and thus the precise ionospheric delay at each frequency.

Military receivers can decode the P(Y) code transmitted on both L1 and L2. Without decryption keys, it is still possible to use a *codeless* technique to compare the P(Y) codes on L1 and L2 to gain much of the same error information. However, this technique is slow, so it is currently available only on specialized surveying equipment. In the future, additional civilian codes are expected to be transmitted on the L2 and L5 frequencies. All users will then be able to perform dual-frequency measurements and directly compute ionospheric delay errors.

A second form of precise monitoring is called *Carrier-Phase Enhancement* (CPGPS). This corrects the error that arises because the pulse transition of the PRN is not instantaneous, and thus the correlation (satellite–receiver sequence matching) operation is imperfect. CPGPS uses the L1 carrier wave, which has a period of $\dfrac{1s}{1575.42 \times 10^6} = 0.63475ns \approx 1ns$, which is about one-thousandth of the C/A Gold code bit period of $\dfrac{1s}{1023 \times 10^3} = 977.5ns \approx 1000ns$, to act as an additional clock signal and resolve the uncertainty. The phase difference error in the normal GPS amounts to 2–3 meters (7–10 ft) of ambiguity. CPGPS working to within 1% of perfect transition reduces this error to 3 centimeters (1.2 in) of ambiguity. By eliminating this error source, CPGPS coupled with DGPS normally realizes between 20–30 centimeters (8–12 in) of absolute accuracy.

Relative Kinematic Positioning (RKP) is a third alternative for a precise GPS-based positioning system. In this approach, determination of range signal can be resolved to a precision of less than 10 centimeters (4 in). This is done by resolving the number of cycles that the signal is transmitted and received by the receiver by using a combination of differential GPS (DGPS) correction data, transmitting GPS signal phase information and ambiguity resolution techniques via statistical tests—possibly with processing in real-time (real-time kinematic positioning, RTK).

Timekeeping

Leap Seconds

While most clocks derive their time from Coordinated Universal Time (UTC), the atomic clocks on the satellites are set to GPS time. The difference is that GPS time is not corrected to match the rotation of the Earth, so it does not contain leap seconds or other corrections that are periodically added to UTC. GPS time was set to match UTC in 1980, but has since diverged. The lack of corrections means that GPS time remains at a constant offset with International Atomic Time (TAI) (TAI – GPS = 19 seconds). Periodic corrections are performed to the on-board clocks to keep them synchronized with ground clocks.

The GPS navigation message includes the difference between GPS time and UTC. As of January 2017, GPS time is 18 seconds ahead of UTC because of the leap second added to UTC on December 31, 2016. Receivers subtract this offset from GPS time to calculate UTC and specific timezone values. New GPS units may not show the correct UTC time until after receiving the UTC offset message. The GPS-UTC offset field can accommodate 255 leap seconds (eight bits).

Accuracy

GPS time is theoretically accurate to about 14 nanoseconds. However, most receivers lose accuracy in the interpretation of the signals and are only accurate to 100 nanoseconds.

Format

As opposed to the year, month, and day format of the Gregorian calendar, the GPS date is expressed as a week number and a seconds-into-week number. The week number is transmitted as a ten-bit field in the C/A and P(Y) navigation messages, and so it becomes zero again every 1,024 weeks (19.6 years). GPS week zero started at 00:00:00 UTC (00:00:19 TAI) on January 6, 1980, and the week number became zero again for the first time at 23:59:47 UTC on August 21,

1999 (00:00:19 TAI on August 22, 1999). To determine the current Gregorian date, a GPS receiver must be provided with the approximate date (to within 3,584 days) to correctly translate the GPS date signal. To address this concern the modernized GPS navigation message uses a 13-bit field that only repeats every 8,192 weeks (157 years), thus lasting until the year 2137 (157 years after GPS week zero).

Carrier Phase Tracking (Surveying)

Another method that is used in surveying applications is carrier phase tracking. The period of the carrier frequency multiplied by the speed of light gives the wavelength, which is about 0.19 meters for the L1 carrier. Accuracy within 1% of wavelength in detecting the leading edge reduces this component of pseudorange error to as little as 2 millimeters. This compares to 3 meters for the C/A code and 0.3 meters for the P code.

However, 2 millimeter accuracy requires measuring the total phase—the number of waves multiplied by the wavelength plus the fractional wavelength, which requires specially equipped receivers. This method has many surveying applications. It is accurate enough for real-time tracking of the very slow motions of tectonic plates, typically 0–100 mm (0–4 inches) per year.

Triple differencing followed by numerical root finding, and a mathematical technique called least squares can estimate the position of one receiver given the position of another. First, compute the difference between satellites, then between receivers, and finally between epochs. Other orders of taking differences are equally valid. Detailed discussion of the errors is omitted.

The satellite carrier total phase can be measured with ambiguity as to the number of cycles. Let $\phi(r_i, s_j, t_k)$ denote the phase of the carrier of satellite j measured by receiver i at time t_k. This notation shows the meaning of the subscripts i, j, and k. The receiver (r), satellite (s), and time (t) come in alphabetical order as arguments of ϕ and to balance readability and conciseness, let $\phi_{i,j,k} = \phi(r_i, s_j, t_k)$ be a concise abbreviation. Also we define three functions, : $\Delta^r, \Delta^s, \Delta^t$, which return differences between receivers, satellites, and time points, respectively. Each function has variables with three subscripts as its arguments. These three functions are defined below. If $\alpha_{i,j,k}$ is a function of the three integer arguments, $i, j,$ and k then it is a valid argument for the functions, : $\Delta^r, \Delta^s, \Delta^t$, with the values defined as

$$\Delta^r(\alpha_{i,j,k}) = \alpha_{i+1,j,k} - \alpha_{i,j,k},$$

$$\Delta^s(\alpha_{i,j,k}) = \alpha_{i,j+1,k} - \alpha_{i,j,k},$$

and

$$\Delta^t(\alpha_{i,j,k}) = \alpha_{i,j,k+1} - \alpha_{i,j,k}.$$

Also if $\alpha_{i,j,k}$ and $\beta_{l,m,n}$ are valid arguments for the three functions and a and b are constants then $(a\,\alpha_{i,j,k} + b\,\beta_{l,m,n})$ is a valid argument with values defined as

$$\Delta^r(a\,\alpha_{i,j,k} + b\,\beta_{l,m,n}) = a\,\Delta^r(\alpha_{i,j,k}) + b\,\Delta^r(\beta_{l,m,n}),$$

$$\Delta^s (a\ \alpha_{i,j,k} + b\ \beta_{l,m,n}) = a\ \Delta^s (\alpha_{i,j,k}) + b\ \Delta^s (\beta_{l,m,n}),$$

and

$$\Delta^t (a\ \alpha_{i,j,k} + b\ \beta_{l,m,n}) = a\ \Delta^t (\alpha_{i,j,k}) + b\ \Delta^t (\beta_{l,m,n}).$$

Receiver clock errors can be approximately eliminated by differencing the phases measured from satellite 1 with that from satellite 2 at the same epoch. This difference is designated as $\Delta^s (\phi_{1,1,1}) = \phi_{1,2,1} - \phi_{1,1,1}$

Double differencing computes the difference of receiver 1's satellite difference from that of receiver 2. This approximately eliminates satellite clock errors. This double difference is:

$$\Delta^r (\Delta^s (\phi_{1,1,1})) = \Delta^r (\phi_{1,2,1} - \phi_{1,1,1}) = \Delta^r (\phi_{1,2,1}) - \Delta^r (\phi_{1,1,1}) = (\phi_{2,2,1} - \phi_{1,2,1}) - (\phi_{2,1,1} - \phi_{1,1,1})$$

Triple differencing subtracts the receiver difference from time 1 from that of time 2. This eliminates the ambiguity associated with the integral number of wavelengths in carrier phase provided this ambiguity does not change with time. Thus the triple difference result eliminates practically all clock bias errors and the integer ambiguity. Atmospheric delay and satellite ephemeris errors have been significantly reduced. This triple difference is:

$$\Delta^t (\Delta^r (\Delta^s (\phi_{1,1,1})))$$

Triple difference results can be used to estimate unknown variables. For example, if the position of receiver 1 is known but the position of receiver 2 unknown, it may be possible to estimate the position of receiver 2 using numerical root finding and least squares. Triple difference results for three independent time pairs may be sufficient to solve for receiver 2's three position components. This may require a numerical procedure. An approximation of receiver 2's position is required to use such a numerical method. This initial value can probably be provided from the navigation message and the intersection of sphere surfaces. Such a reasonable estimate can be key to successful multidimensional root finding. Iterating from three time pairs and a fairly good initial value produces one observed triple difference result for receiver 2's position. Processing additional time pairs can improve accuracy, overdetermining the answer with multiple solutions. Least squares can estimate an overdetermined system. Least squares determines the position of receiver 2 that best fits the observed triple difference results for receiver 2 positions under the criterion of minimizing the sum of the squares.

Regulatory Spectrum Issues Concerning GPS Receivers

In the United States, GPS receivers are regulated under the Federal Communications Commission's (FCC) Part 15 rules. As indicated in the manuals of GPS-enabled devices sold in the United States, as a Part 15 device, it "must accept any interference received, including interference that may cause undesired operation." With respect to GPS devices in particular, the FCC states that GPS receiver manufacturers, "must use receivers that reasonably discriminate against reception of signals outside their allocated spectrum." For the last 30 years, GPS receivers have operated next to the Mobile Satellite Service band, and have discriminated against reception of mobile satellite services, such as Inmarsat, without any issue.

The spectrum allocated for GPS L1 use by the FCC is 1559 to 1610 MHz, while the spectrum allocated for satellite-to-ground use owned by Lightsquared is the Mobile Satellite Service band. Since 1996, the FCC has authorized licensed use of the spectrum neighboring the GPS band of 1525 to 1559 MHz to the Virginia company LightSquared. On March 1, 2001, the FCC received an application from LightSquared's predecessor, Motient Services, to use their allocated frequencies for an integrated satellite-terrestrial service. In 2002, the U.S. GPS Industry Council came to an out-of-band-emissions (OOBE) agreement with LightSquared to prevent transmissions from LightSquared's ground-based stations from emitting transmissions into the neighboring GPS band of 1559 to 1610 MHz. In 2004, the FCC adopted the OOBE agreement in its authorization for LightSquared to deploy a ground-based network ancillary to their satellite system – known as the Ancillary Tower Components (ATCs) – "We will authorize MSS ATC subject to conditions that ensure that the added terrestrial component remains ancillary to the principal MSS offering. We do not intend, nor will we permit, the terrestrial component to become a stand-alone service." This authorization was reviewed and approved by the U.S. Interdepartment Radio Advisory Committee, which includes the U.S. Department of Agriculture, U.S. Air Force, U.S. Army, U.S. Coast Guard, Federal Aviation Administration, National Aeronautics and Space Administration, Interior, and U.S. Department of Transportation.

In January 2011, the FCC conditionally authorized LightSquared's wholesale customers—such as Best Buy, Sharp, and C Spire—to only purchase an integrated satellite-ground-based service from LightSquared and re-sell that integrated service on devices that are equipped to only use the ground-based signal using LightSquared's allocated frequencies of 1525 to 1559 MHz. In December 2010, GPS receiver manufacturers expressed concerns to the FCC that LightSquared's signal would interfere with GPS receiver devices although the FCC's policy considerations leading up to the January 2011 order did not pertain to any proposed changes to the maximum number of ground-based LightSquared stations or the maximum power at which these stations could operate. The January 2011 order makes final authorization contingent upon studies of GPS interference issues carried out by a LightSquared led working group along with GPS industry and Federal agency participation. On February 14, 2012, the FCC initiated proceedings to vacate LightSquared's Conditional Waiver Order based on the NTIA's conclusion that there was currently no practical way to mitigate potential GPS interference.

GPS receiver manufacturers design GPS receivers to use spectrum beyond the GPS-allocated band. In some cases, GPS receivers are designed to use up to 400 MHz of spectrum in either direction of the L1 frequency of 1575.42 MHz, because mobile satellite services in those regions are broadcasting from space to ground, and at power levels commensurate with mobile satellite services. However, as regulated under the FCC's Part 15 rules, GPS receivers are not warranted protection from signals outside GPS-allocated spectrum. This is why GPS operates next to the Mobile Satellite Service band, and also why the Mobile Satellite Service band operates next to GPS. The symbiotic relationship of spectrum allocation ensures that users of both bands are able to operate cooperatively and freely.

The FCC adopted rules in February 2003 that allowed Mobile Satellite Service (MSS) licensees such as LightSquared to construct a small number of ancillary ground-based towers in their licensed spectrum to "promote more efficient use of terrestrial wireless spectrum." In those 2003 rules, the FCC stated "As a preliminary matter, terrestrial [Commercial Mobile Radio Service ("CMRS")] and MSS ATC are expected to have different prices, coverage, product acceptance and distribution; therefore, the two services appear, at best, to be imperfect substitutes for one another that would be operating in predominately different market segments... MSS ATC is unlikely to compete directly with terrestrial CMRS for the same customer base...". In 2004, the FCC clarified that the ground-based towers would be ancillary, noting that "We will authorize MSS ATC subject to conditions that ensure that the added terrestrial component remains ancillary to the principal MSS offering. We do not intend, nor will we permit, the terrestrial component to become a stand-alone service." In July 2010, the FCC stated that it expected LightSquared to use its authority to offer an integrated satellite-terrestrial service to "provide mobile broadband services similar to those provided by terrestrial mobile providers and enhance competition in the mobile broadband sector." However, GPS receiver manufacturers have argued that LightSquared's licensed spectrum of 1525 to 1559 MHz was never envisioned as being used for high-speed wireless broadband based on the 2003 and 2004 FCC ATC rulings making clear that the Ancillary Tower Component (ATC) would be, in fact, ancillary to the primary satellite component. To build public support of efforts to continue the 2004 FCC authorization of LightSquared's ancillary terrestrial component vs. a simple ground-based LTE service in the Mobile Satellite Service band, GPS receiver manufacturer Trimble Navigation Ltd. formed the "Coalition To Save Our GPS."

The FCC and LightSquared have each made public commitments to solve the GPS interference issue before the network is allowed to operate. However, according to Chris Dancy of the Aircraft Owners and Pilots Association, airline pilots with the type of systems that would be affected "may go off course and not even realize it." The problems could also affect the Federal Aviation Administration upgrade to the air traffic control system, United States Defense Department guidance, and local emergency services including 911.

On February 14, 2012, the U.S. Federal Communications Commission (FCC) moved to bar LightSquared's planned national broadband network after being informed by the National Telecommunications and Information Administration (NTIA), the federal agency that coordinates spectrum uses for the military and other federal government entities, that "there is no practical way to mitigate potential interference at this time". LightSquared is challenging the FCC's action.

Other Systems

Other satellite navigation systems in use or various states of development include:

- GLONASS – Russia's global navigation system. Fully operational worldwide.
- Galileo – a global system being developed by the European Union and other partner countries, which began operation in 2016, and is expected to be fully deployed by 2020.
- Beidou – People's Republic of China's regional system, currently limited to Asia and the West Pacific, global coverage planned to be operational by 2020
- IRNSS - A regional navigation system developed by the Indian Space Research Organisation.
- QZSS - A regional navigation system in development that would be receivable within Japan.

Lidar

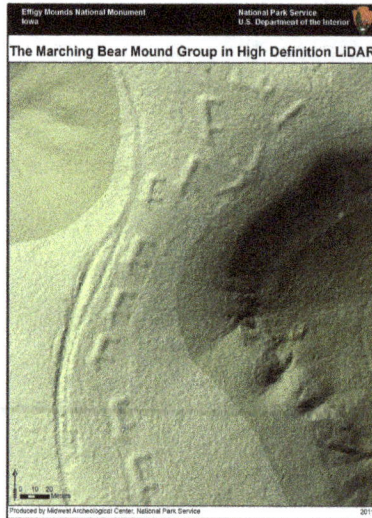

Lidar-derived image of Marching Bears Mound Group, Effigy Mounds National Monument

Lidar (also called LIDAR, LiDAR, and LADAR) is a surveying method that measures distance to a target by illuminating that target with a pulsed laser light, and measuring the reflected pulses with a sensor. Differences in laser return times and wavelengths can then be used to make digital 3D-representations of the target. The name *lidar*, sometimes considered an acronym of *Light Detection And Ranging* (sometimes *Light Imaging, Detection, And Ranging*), was originally a portmanteau of *light* and *radar*.

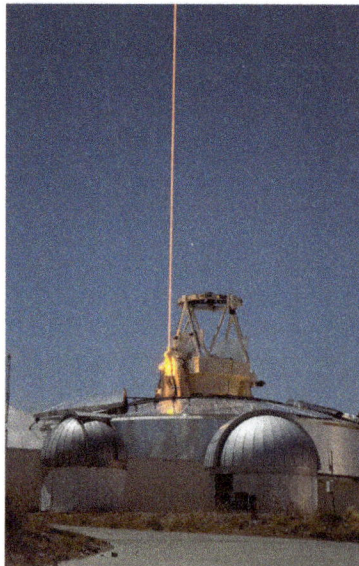

A FASOR used at the Starfire Optical Range for lidar and laser guide star experiments is tuned to the sodium D2a line and used to excite sodium atoms in the upper atmosphere.

Lidar is popularly used to make high-resolution maps, with applications in geodesy, geomatics, archaeology, geography, geology, geomorphology, seismology, forestry, atmospheric physics, laser guidance, airborne laser swath mapping (ALSM), and laser altimetry. The technology is also used for control and navigation for some autonomous cars. Lidar sometimes is called *laser scanning* and *3D scanning*, with terrestrial, airborne, and mobile applications.

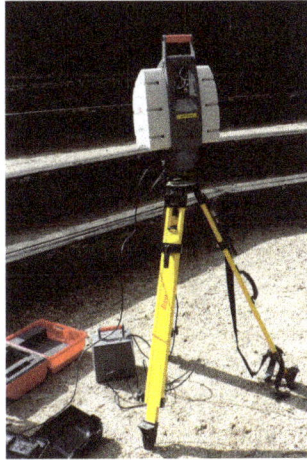

This lidar may be used to scan buildings, rock formations, etc., to produce a 3D model.
The lidar can aim its laser beam in a wide range: its head rotates horizontally; a mirror tilts vertically.
The laser beam is used to measure the distance to the first object on its path.

History and Etymology

Lidar originated in the early 1960s, shortly after the invention of the laser, and combined laser-focused imaging with the ability to calculate distances by measuring the time for a signal to return using appropriate sensors and data acquisition electronics. Its first applications came in meteorology, where the National Center for Atmospheric Research used it to measure clouds. The general public became aware of the accuracy and usefulness of lidar systems in 1971 during the Apollo 15 mission, when astronauts used a laser altimeter to map the surface of the moon.

Although some sources treat the word "lidar" as an acronym, the term originated as a portmanteau of "light" and "radar". The first published mention of lidar, in 1963, makes this clear: "Eventually the laser may provide an extremely sensitive detector of particular wavelengths from distant objects. Meanwhile, it is being used to study the moon by 'lidar' (light radar) ..." The *Oxford English Dictionary* supports this etymology.

The interpretation of "lidar" as an acronym ("LIDAR" or "LiDAR") came later, beginning in 1970, based on the assumption that since the base term "radar" originally started as an acronym for "RAdio Detection And Ranging", "LIDAR" must stand for "LIght Detection And Ranging", or for "Laser Imaging, Detection and Ranging". Although the English language no longer treats "radar" as an acronym and printed texts universally present the word uncapitalized, the word "lidar" became capitalized as "LIDAR" or "LiDAR" in some publications beginning in the 1980s. Currently no consensus exists on capitalization, reflecting uncertainty about whether or not "lidar" is an acronym, and if it is an acronym, whether it should appear in lower case, like "radar". Various publications refer to lidar as "LIDAR", "LiDAR", "LIDaR", or "Lidar". The USGS uses both "LIDAR" and "lidar", sometimes in the same document; the *New York Times* predominantly uses "lidar" for staff written articles, although contributing news feeds such as Reuters may use Lidar; and Uber sometimes uses Lidar for public statements.

General Description

Lidar uses ultraviolet, visible, or near infrared light to image objects. It can target a wide range of materials, including non-metallic objects, rocks, rain, chemical compounds, aerosols, clouds and even single molecules. A narrow laser-beam can map physical features with very high resolutions; for example, an aircraft can map terrain at 30-centimetre (12 in) resolution or better.

Lidar has been used extensively for atmospheric research and meteorology. Lidar instruments fitted to aircraft and satellites carry out surveying and mapping – a recent example being the U.S. Geological Survey Experimental Advanced Airborne Research Lidar. NASA has identified lidar as a key technology for enabling autonomous precision safe landing of future robotic and crewed lunar-landing vehicles.

Wavelengths vary to suit the target: from about 10 micrometers to the UV (approximately 250 nm). Typically light is reflected via backscattering, as opposed to pure reflection one might find with a mirror. Different types of scattering are used for different lidar applications: most commonly Rayleigh scattering, Mie scattering, Raman scattering, and fluorescence. Based on different kinds of backscattering, the lidar can be accordingly called Rayleigh lidar, Mie lidar, Raman lidar, Na/Fe/K Fluorescence lidar, and so on. Suitable combinations of wavelengths can allow for remote mapping of atmospheric contents by identifying wavelength-dependent changes in the intensity of the returned signal.

Design

In general there are two kinds of lidar detection schemes: "incoherent" or direct energy detection (which principally measures amplitude changes of the reflected light) and coherent detection (which is best for measuring Doppler shifts, or changes in phase of the reflected light). Coherent systems generally use optical heterodyne detection, which, being more sensitive than direct detection, allows them to operate at a much lower power but at the expense of more complex transceiver requirements.

In both coherent and incoherent lidar, there are two types of pulse models: *micropulse lidar* systems and *high energy* systems. Micropulse systems utilizing intermittent bursts of energy have developed as a result of the ever-increasing amount of computer power available combined with advances in laser technology. They use considerably less energy in the laser, typically on the order of one microjoule, and are often "eye-safe," meaning they can be used without safety precautions. High-power systems are common in atmospheric research, where they are widely used for measuring many atmospheric parameters: the height, layering and densities of clouds, cloud particle properties (extinction coefficient, backscatter coefficient, depolarization), temperature, pressure, wind, humidity, and trace gas concentration (ozone, methane, nitrous oxide, etc.).

There are several major components to a lidar system:

1. Laser — 600–1000 nm lasers are most common for non-scientific applications. They are inexpensive, but since they can be focused and easily absorbed by the eye, the maximum power is limited by the need to make them eye-safe. Eye-safety is often a requirement for most applications. A common alternative, 1550 nm lasers, are eye-safe at much higher

power levels since this wavelength is not focused by the eye, but the detector technology is less advanced and so these wavelengths are generally used at longer ranges with lower accuracies. They are also used for military applications as 1550 nm is not visible in night vision goggles, unlike the shorter 1000 nm infrared laser. Airborne topographic mapping lidars generally use 1064 nm diode pumped YAG lasers, while bathymetric (underwater depth research) systems generally use 532 nm frequency doubled diode pumped YAG lasers because 532 nm penetrates water with much less attenuation than does 1064 nm. Laser settings include the laser repetition rate (which controls the data collection speed). Pulse length is generally an attribute of the laser cavity length, the number of passes required through the gain material (YAG, YLF, etc.), and Q-switch (pulsing) speed. Better target resolution is achieved with shorter pulses, provided the lidar receiver detectors and electronics have sufficient bandwidth.

2. Scanner and optics — How fast images can be developed is also affected by the speed at which they are scanned. There are several options to scan the azimuth and elevation, including dual oscillating plane mirrors, a combination with a polygon mirror and a dual axis scanner. Optic choices affect the angular resolution and range that can be detected. A hole mirror or a beam splitter are options to collect a return signal.

3. Photodetector and receiver electronics — Two main photodetector technologies are used in lidars: solid state photodetectors, such as silicon avalanche photodiodes, or photomultipliers. The sensitivity of the receiver is another parameter that has to be balanced in a lidar design.

4. Position and navigation systems — Lidar sensors that are mounted on mobile platforms such as airplanes or satellites require instrumentation to determine the absolute position and orientation of the sensor. Such devices generally include a Global Positioning System receiver and an Inertial Measurement Unit (IMU).

3D imaging can be achieved using both scanning and non-scanning systems. "3D gated viewing laser radar" is a non-scanning laser ranging system that applies a pulsed laser and a fast gated camera. Research has begun for virtual beam steering using Digital Light Processing (DLP) technology.

Imaging lidar can also be performed using arrays of high speed detectors and modulation sensitive detector arrays typically built on single chips using Complementary metal–oxide–semiconductor (CMOS) and hybrid CMOS/Charge-coupled device (CCD) fabrication techniques. In these devices each pixel performs some local processing such as demodulation or gating at high speed, down-converting the signals to video rate so that the array may be read like a camera. Using this technique many thousands of pixels / channels may be acquired simultaneously. High resolution 3D lidar cameras use homodyne detection with an electronic CCD or CMOS shutter.

A coherent imaging lidar uses synthetic array heterodyne detection to enable a staring single element receiver to act as though it were an imaging array.

In 2014 Lincoln Laboratory announced a new imaging chip with more than 16,384 pixels, each able to image a single photon, enabling them to capture a wide area in a single image. An earlier generation of the technology with one-quarter as many pixels was dispatched by the U.S. military after the January 2010 Haiti earthquake; a single pass by a business jet at 3,000 meters (10,000 ft.) over Port-au-Prince was able to capture instantaneous snapshots of 600-meter squares of the city at a resolution of 30 centimetres (12 in), displaying the precise height of rubble strewn in city streets. The new system is another 10x faster. The chip uses indium gallium arsenide (InGaAs), which operates in the infrared spectrum at a relatively long wavelength that allows for higher power and longer ranges. In many applications, such as self-driving cars, the new system will lower costs by not requiring a mechanical component to aim the chip. InGaAs uses less hazardous wavelengths than conventional silicon detectors, which operate at visual wavelengths.

Sensors

As opposed to passive sensors that detect energy naturally emitted from an object, lidar uses active sensors, which emit their own energy source for illumination. The energy source hits objects and the reflected energy from the objects is detected and measured by sensors. Active sensors used in the microwave portion of the electromagnetic spectrum. Lidar is an example of active sensor and uses a laser (light amplification by stimulated emission of radiation) radar to transmit a light pulse and a receiver with sensitive detectors to measure the backscattered or reflected light. Distance to the object is determined by recording the time between transmitted and backscattered pulses and by using the speed of light to calculate the distance traveled

Types of Applications

Lidar has a wide range of applications which can be divided into airborne and terrestrial types. These different types of applications require scanners with varying specifications based on the data's purpose, the size of the area to be captured, the range of measurement desired, the cost of equipment, and more.

Airborne Lidar

Airborne lidar (also *airborne laser scanning*) is when a laser scanner, while attached to a plane during flight, creates a 3D point cloud model of the landscape. This is currently the most detailed and accurate method of creating digital elevation models, replacing photogrammetry. One major advantage in comparison with photogrammetry is the ability to filter out reflections from vegetation from the point cloud model to create a digital surface model which represents ground surfaces such as rivers, paths, cultural heritage sites, etc., which are concealed by trees. Within the category of airborne lidar, there is sometimes a distinction made between high-altitude and low-altitude applications, but the main difference is a reduction in both accuracy and point density of data acquired at higher altitudes. Airborne lidar can also be used to create bathymetric models in shallow water.

The main constituents of airborne lidar include digital elevation models (DEM) and digital survey models (DSM). The points and ground points are the vectors of discrete points while DEM and DSM are interpolated raster grids of discrete points. The process also involves capturing of digital aerial photographs. In order to interpret deep seated landslides for example, under the cover of vegetation, scarps, tension cracks or tipped trees air borne lidar is used. Air borne lidar digital elevation models can see through the canopy of forest cover, perform detailed measurements of scarps, erosion and tilting of electric poles.

Airborne lidar data is processed using a toolbox called Toolbox for Lidar Data Filtering and Forest Studies (TIFFS) for lidar data filtering and terrain study software. The data is interpolated to digital terrain models using the software. The laser is directed at the region to be mapped and each point's height above the ground is calculated by subtracting the original z-coordinate from the corresponding digital terrain model elevation. Based on this height above the ground the non-vegetation data is obtained which may include objects such as buildings, electric power lines, flying birds etc. The rest of the points are treated as vegetation and used for modeling and mapping. Within each of these plots, lidar metrics are calculated by calculating statistics such as mean, standard deviation, skewness, percentiles, quadratic mean etc.

Airborne Lidar Bathymetric Technology

The airborne lidar bathymetric technological system involves the measurement of time of flight of a signal from a source to its return to the sensor. The data acquisition technique involves a sea floor mapping component and a ground truth component that includes video transects and sampling. It works using a green spectrum (532 nm) laser beam. Two beams are projected onto a fast rotating mirror, which creates an array of points. One of the beams penetrates the water and also detects the bottom surface of the water under favorable conditions.

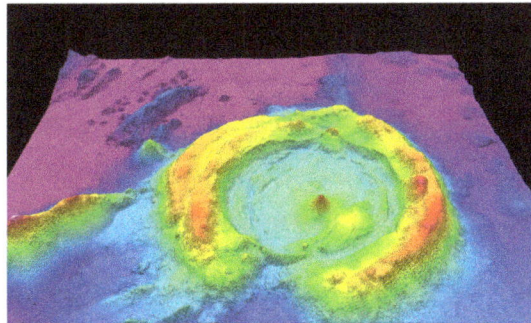

Airborne Lidar Bathymetric Technology-High-resolution multibeam lidar map showing spectacularly faulted and deformed seafloor geology, in shaded relief and coloured by depth.

The data obtained shows the full extent of the land surface exposed above the sea floor. This technique is extremely useful as it will play an important role in the major sea floor mapping program. The mapping yields onshore topography as well as under water elevations. Sea floor reflectance imaging is another solution product from this system which can benefit mapping of underwater habitats. This technique has been used for three dimensional image mapping of California's waters using a hydrographic lidar.

LIDAR scanning performed with the OnyxStar FOX-C8 HD UAV from AltiGator

Drones are now being used with laser scanners, as well as other remote sensors, as a more economical method to scan smaller areas. The possibility of drone remote sensing also eliminates any danger that crews of a manned aircraft may be subjected to in difficult terrain or remote areas.

Terrestrial Lidar

Terrestrial applications of lidar (also *terrestrial laser scanning*) happen on the Earth's surface and can be both stationary or mobile. Stationary terrestrial scanning is most common as a survey method, for example in conventional topography, monitoring, cultural heritage documentation and forensics. The 3D point clouds acquired from these types of scanners can be matched with digital images taken of the scanned area from the scanner's location to create realistic looking 3D models in a relatively short time when compared to other technologies. Each point in the point cloud is given the colour of the pixel from the image taken located at the same angle as the laser beam that created the point.

Mobile lidar (also *mobile laser scanning*) is when two or more scanners are attached to a moving vehicle to collect data along a path. These scanners are almost always paired with other kinds of equipment, including GNSS receivers and IMUs. One example application is surveying streets, where power lines, exact bridge heights, bordering trees, etc. all need to be taken into account. Instead of collecting each of these measurements individually in the field with a tachymeter, a 3D model from a point cloud can be created where all of the measurements needed can be made, depending on the quality of the data collected. This eliminates the problem of forgetting to take a measurement, so long as the model is available, reliable and has an appropriate level of accuracy.

Terrestrial lidar mapping involves a process of occupancy grid map generation. The process involves an array of cells divided into grids which employs a process to store the height values when lidar data falls into the respective grid cell. A binary map is then created by applying a particular threshold to the cell values for further processing. The next step is to process the radial distance and z-coordinates from each scan to identify which 3D points correspond to each of the specified grid cell leading to the process of data formation.

Data Formation

For the autonomous applications a lidar with a 3D scanner is used for obstacle detection and collision avoidance.The scanner measures radial distance at different angular resolutions. The lidar emits a single laser ray and uses an interior rotating mirror to distribute the laser ray covering a large field of view.

Object Detection

The object detection procedure consists of three steps: Laser point feature calculation and prior filtering, 3D segmentation, object classification and 2D position calculation. The features for segmentation are calculated in the vicinity of a fixed radial distance in 2D and 3D.These include:

Amplitude Density

This involves the percentage of points in the vicinity with amplitude values below a particular threshold. Points with a density lower than the threshold are removed from segmentation and further processing.

Height Above Local Minimum in 2D Search Radius

In this case the points which have a height lower than a certain threshold are removed. On segmentation of the point cloud, all the segments satisfying the object based criteria are included in individual objects.

Advantages

The main advantages of object detection and data formation are that high point density and high spatial resolution is obtained. Also the accuracy of the acquired data sets is increased to a considerable extent.

Disadvantages

Although significant amount of research has been done for object classification from 3D point clouds, the direct extraction of individual points from lidar inputs has not been achieved.

Object Detection for Transportation Systems

In transportation systems, to ensure vehicle and passenger safety and to develop electronic systems that deliver driver assistance, understanding vehicle and surrounding environment is essential. Lidar system plays an important role in safety of transportation systems. Lots of electronic systems which add to the driver assistance and vehicle safety such as Adaptive Cruise Control (ACC), Emergency Brake Assist, Anti-lock Braking System (ABS) depends on the detection of vehicle environment to act autonomously or semi-autonomously. Lidar mapping and estimation achieve this.

Basics overview: Current lidar systems use rotating hexagonal mirrors which split the laser beam. The upper three beams are used for vehicle and obstacles ahead and the lower beams are used to detect lane markings and road features. The major advantage of using lidar is that the spatial structure is obtained and this data can be fused with other sensors like RADAR etc. to get a better picture of the vehicle environment in terms of static and dynamic properties of the objects present in the environment.

Applications

This lidar-equipped mobile robot uses its lidar to construct a map and avoid obstacles.

There are a wide variety of applications for lidar, in addition to the applications listed below, as it is often mentioned in National lidar dataset programs.

Agriculture

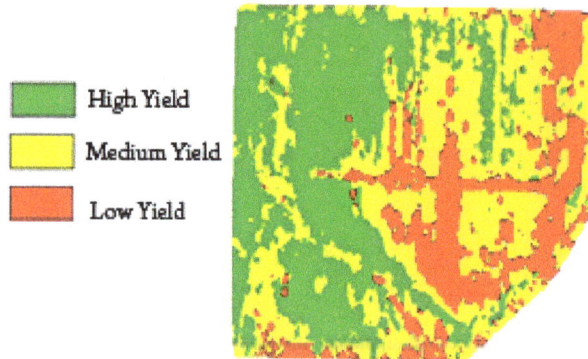

Agricultural Research Service scientists have developed a way to incorporate lidar with yield rates on agricultural fields. This technology will help farmers improve their yields by directing their resources toward the high-yield sections of their land.

Lidar also can be used to help farmers determine which areas of their fields to apply costly fertilizer. Lidar can create a topographical map of the fields and reveals the slopes and sun exposure of the farm land. Researchers at the Agricultural Research Service blended this topographical information with the farmland yield results from previous years. From this information, researchers categorized the farm land into high-, medium-, or low-yield zones. This technology is valuable to farmers because it indicates which areas to apply the expensive fertilizers to achieve the highest crop yield.

Another application of lidar beyond crop health and terrain mapping is crop mapping in orchards and vineyards. Vehicles equipped with lidar sensors can detect foliage growth to determine if pruning or other maintenance needs to take place, detect variations in fruit production, or perform automated tree counts.

Lidar is useful in GPS-denied situations, such as in nut and fruit orchards where GPS signals to farm equipment featuring precision agriculture technology or a driverless tractor may be partially or completely blocked by overhanging foliage. Lidar sensors can detect the edges of rows so that farming equipment can continue moving until GPS signal can be reestablished.

Plant Species Classification Using a 3D Lidar Sensor and Machine Learning

This is a methodology for differentiating plant species using a 3D lidar sensor and machine learning. The procedure involves the application of logistic regression functions, support vector machines and neural networks.

Problem Statement

Autonomous robots have been used for a variety of purposes in agriculture ranging from seed and fertilizer dispersions, sensing techniques as well as crop scouting for the task of weed control. In order to execute weed control one of the major factors is the detection and classification of the plant and species. In addition it is difficult to differentiate the characteristics and appearances of the plant and convert the data into computer understandable form. Also, it is not possible to accurately detect structures of the plant in 3D data. The problem is solved by classifying the plant species by using a set of example plants and machine learning methods.

Algorithm Overview

The input to the system are 3D point clouds with range and reflectance values from a 3D laser sensor. The species of the plant is known by applying preprocessing techniques to provide the point cloud for further processing. If the species is already known then the extracted features are added as new data. The particular plant species is labeled and its features are initially stored as an example to identify the particular plant species in the real environment.

Pre-processing

In this process, there are three steps. The first step is to identify the ground plane and the points returning from the ground plane are eliminated. In the second step the remaining points are grouped as one and the individual points are detected. The third step involves the identification of the plant species. An individual coordinate system is developed for each plant. The x and y coordinates are placed on the ground while the z-axis is along the center of the plant. In this way individual plants in row cultivations are separated. The next step is to identify the plant space.

Feature Extraction

To understand a set of features to identify the appearance of the plant species the criteria of size and rotation invariance of the 3D laser sensor is used. The set of features are divided into two groups: Reflectance and geometrical features

Reflectance Features

Plant species have different variations based on their reflectance values. It is an important tool for plant classification. The reflectance value encodes information about the leaf size and shape. For each of the measuring points a reflectance value between 0 and 1 are determined.

Geometrical Features

The laser sensor also provides a range of data which can be transformed into 3D coordinates. Using the coordinates, the shape, dimension and structure of the plant can be characterized.

Classification

Classification of the plants is based on machine learning. For the learning process a training set is generated with known samples. Machine learning toolbox WEKA is used for this purpose.

Tested Plants

The training set is generated with the help of potted plants available in common nurseries. Different plant species is chosen and the corresponding plant cloud is developed. Besides the plant the ground coordinate system is also sensed to determine the plant coordinate system that is essential for feature calculation.

Advantages

This method for distinguishing plant species is highly efficient since it uses a low resolution 3D lidar sensor and supervised learning. It includes easy to compute feature set with common statistical features which are independent of the plant size.

Archaeology

Lidar has many applications in the field of archaeology including aiding in the planning of field campaigns, mapping features beneath forest canopy, and providing an overview of broad, continuous features that may be indistinguishable on the ground. Lidar can also provide archaeologists with the ability to create high-resolution digital elevation models (DEMs) of archaeological sites that can reveal micro-topography that are otherwise hidden by vegetation. Lidar-derived products can be easily integrated into a Geographic Information System (GIS) for analysis and interpretation. For example, at Fort Beauséjour – Fort Cumberland National Historic Site, Canada, previously undiscovered archaeological features below forest canopy have been mapped that are related to the siege of the Fort in 1755. Features that could not be distinguished on the ground or through aerial photography were identified by overlaying hillshades of the DEM created with artificial illumination from various angles. With lidar, the ability to produce high-resolution datasets quickly and relatively cheaply can be an advantage. Beyond efficiency, its ability to penetrate forest canopy has led to the discovery of features that were not distinguishable through traditional geo-spatial methods and are difficult to reach through field surveys, as in work at Caracol by Arlen Chase and his wife Diane Zaino Chase. The intensity of the returned signal can be used to detect features buried under flat vegetated surfaces such as fields, especially when mapping using the infrared spectrum. The presence of these features affects plant growth and thus the amount of infrared light reflected back. In 2012, lidar was used by a team attempting to find the legendary city of La Ciudad Blanca in the Honduran jungle. During a seven-day mapping period, they found evidence of extensive man-made structures. In June 2013 the rediscovery of the city of Mahendraparvata was announced. In another study, lidar was used to reveal stone walls, building foundations, abandoned roads, and other features of the landscape in southern New England, USA that had been obscured in aerial photography by the region's dense forest canopy. In May 2012, lidar was used to locate a previously unknown ruined city in the La Mosquitia region of Honduras. In Cambodia, lidar data were used by Demian Evans and Roland Fletcher to reveal anthropogenic changes to Angkor landscape

Autonomous Vehicles

Autonomous vehicles use lidar for obstacle detection and avoidance to navigate safely through environments, using rotating laser beams. Cost map or point cloud outputs from the lidar sensor provide the necessary data for robot software to determine where potential obstacles exist in the environment and where the robot is in relation to those potential obstacles. Singapore's *Singapore-MIT Alliance for Research and Technology (SMART)* is actively developing technologies for autonomous lidar vehicles. Examples of companies that produce lidar sensors commonly used in robotics or vehicle automation are Sick and Hokuyo. Examples of obstacle detection and avoidance products that leverage lidar sensors are the Autonomous Solution, Inc. Forecast 3D Laser System and Velodyne HDL-64E.

Uber self driving car with lidar system on roof

Forecast 3D Laser System using a SICK LMC lidar sensor

The very first generations of automotive adaptive cruise control systems used only lidar sensors.

Approaches of Processing Lidar Data

Below mentioned are various approaches of processing lidar data and utilizing it along with data from other sensors through sensor fusion to detect the vehicle environment conditions.

Lidar image shows road contour, elevation and roadside vegetation.

GRID Based Processing using 3D Lidar and Fusion with RADAR Measurement

In this method, proposed by Philipp Lindner and Gerd Wanielik, laser data is processed using a multidimensional occupancy grid. Data from a 4 layer laser is pre-processed at the signal level and then processed at a higher level to extract the features of the obstacles. A combination 2- and 3-dimensional grid structure is utilized and the space in these structures is tessellated into several discrete cells. This method allows a huge amount of raw measurement data to be effectively handled by collecting it in spatial containers, the cells of the evidence grid. Each cell is associated with a probability measure that identifies the cell occupation. This probability is calculated by using the range measurement of the lidar sensor obtained over time and a new range measurement, which are related using Bayes' theorem. A two dimensional grid can observe an obstacle in front of it, but cannot observe the space behind the obstacle. TO address this, the unknown state behind the obstacle is assigned a probability of 0.5. By introducing the third dimension or in other terms using a multi-layer laser, the spatial configuration of an object could be mapped into the grid structure to a degree of complexity. This is achieved by transferring the measurement points into a 3 dimensional grid. The grid cells which are occupied will possess a probability greater than 0.5 and the mapping would be color coded based on the probability. The cells which are not occupied will possess a probability less than 0.5 and this area will usually be white space. This measurement is then transformed to a grid coordinate system by using the sensor position on the vehicle and the vehicle position in the world coordinate system. The coordinates of the sensor depends upon its location on the vehicle and the coordinates of the vehicle is computed using egomotion estimation, which is estimating the vehicle motion relative to a rigid scene. For this method, the grid profile must be defined. The grid cells touched by the transmitted laser beam are calculated by applying Bresenham's line algorithm. To obtain the spatial extended structure, a connected component analysis of these cells is performed. This information is then passed on to a rotating caliper algorithm to obtain the spatial characteristics of the object. In addition to the lidar detection, RADAR data obtained by using two short range radars is integrated to get additional dynamic properties of the object, such as its velocity. The measurements are assigned to the object using a potential distance function.

Advantages and Disadvantages

The geometric features of the objects are extracted efficiently, from the measurements obtained by the 3D occupancy grid, using rotating caliper algorithm. Fusing the radar data to the lidar measurements give information about the dynamic properties of the obstacle such as velocity and location of the obstacle with respect to the sensor location which helps the vehicle or the driver decide the action to be performed in order to ensure safety. The only concern is the computational requirement to implement this data processing technique. It can be implemented in real time and has been proven efficient if the 3D occupancy grid size is considerably restricted. But this can be improved to an even wider range by using dedicated spatial datastructures that manipulate the spatial data more effectively, for the 3D grid representation.

Fusion of 3D Lidar and Color Camera for Multiple Object Detection and Tracking

The framework proposed in this method by Soonmin Hwang et al., is split into four steps. First, the data from the camera and 3D lidar is input into to the system. Both inputs from lidar and camera are parallelly obtained and the color image from the camera is calibrated with the lidar. To improve the efficiency, horizontal 3D point sampling is applied as pre-processing. Second, the segmentation stage is where the entire 3D points are divided into several groups per the distance from the sensor and local planes from close plane to far plane are sequentially estimated. The local planes are estimated using statistical analysis. The group of points closer to the sensor are used to compute the initial plane. By using the current local plane, the next local plane is estimated by iterative update. The object proposals in the 2D image are used to separate foreground objects from background. For faster and accurate detection and tracking Binarized Normed Gradients for Objectness Estimation at 300fps is used. BING is a combination of normed gradient and its binarized version which speeds up the feature extraction and testing process, to estimate the objectness of an image window. This way the foreground and background objects are separated. To form objects after estimating the objectness of an image using BING, the 3D points are grouped or clustered. Clustering is done using DBSCAN (Density-Based Spatial Clustering of Applications with Noise) algorithm which could be robust due to its less-parametric characteristic. Using the clustered 3D points, i.e. 3D segment, more accurate region-of-interests (RoIs) are generated by projecting 3D points on the 2D image. The third step is detection, which is broadly divided into two parts. First is object detection in 2D image which is achieved using Fast R-CNN as this method doesn't need training and it also considers an image and several regions of interest. Second is object detection in 3D space which is done by using the spin image method. This method extracts local and global histograms to represent a certain object. To merge the results of 2D image and 3D space object detection, same 3D region is considered and two independent classifiers from 2D image and 3D space are applied to the considered region. Scores calibration is done to get a single confidence score from both detectors. This single score is obtained in the form of probability. The final step is tracking. This is done by associating moving objects in present and past frame. For object tracking, segment matching is adopted. Features such as mean, standard deviation, quantized color histograms, volume size and number of 3D points of a segment are computed. Euclidean distance is used to measure differences between segments. To judge the appearance and disappearance of an object, similar segments (obtained based on the Euclidean distance) from two different frames are taken and the physical distance and dissimilarity scores are calculated. If the scores go beyond a range for every segment in previous frame, the object being tracked is considered to have disappeared.

Advantages and Disadvantages

The advantages of this method are using 2D image and 3D data together, F 1-score (which gives a measure of test's accuracy), average precision (AP) are higher than that when only 3D data from lidar is used. These scores are conventional measurements which judge the framework. The drawback of this method is the usage of BING for object proposal estimation as BING predicts a small set of object bounding boxes.

Obstacle Detection and Road Environment Recognition Using Lidar

This method proposed by Kun Zhou et al. not only focuses on object detection and tracking but also recognizes lane marking and road features. As mentioned earlier the lidar systems use rotating hexagonal mirrors which split the laser beam into 6 beams. The upper 3 layers are used to detect the forward objects such as vehicles and roadside objects. The sensor is made of weather-resistant material. The data detected by lidar are clustered to several segments and tracked by Kalman filter. Data clustering here is done based on characteristics of each segment based on object model, which distinguish different objects such as vehicles, signboards etc. These characteristics include the dimensions of the object etc. The reflectors on the rear edges of vehicles are used to differentiate vehicles from other objects. Object tracking is done using a 2-stage Kalman filter considering the stability of tracking and the accelerated motion of objects Lidar reflective intensity data is also used for curb detection by making use of robust regression to deal with occlusions. The road marking is detected using a modified Otsu method by distinguishing rough and shiny surfaces.

Advantages

The advantages of this method Roadside reflectors which indicate lane border are sometimes hidden due to various reasons. Therefore, other information is needed to recognize the road border. The lidar used in this method can measure the reflectivity from the object. Hence, with this data road border can also be recognized. Also the usage of sensor with weather-robust head helps detecting the objects even in bad weather conditions. Canopy Height Model before and after flood is a good example. Lidar can detect high detailed canopy height data as well as its road border.

Advantages of Using Lidar Measurement

Lidar measurements help identify the spatial structure of the obstacle. This helps distinguish objects based on size and estimate the impact of driving over it.

Lidar systems provides better range and a large field of view which helps detecting obstacles on the curves. This is one major advantage over RADAR systems which have a narrower field of view. The fusion of lidar measurement with different sensors makes the system robust and useful in real-time applications, since lidar dependent systems can't estimate the dynamic information about the detected object.

It has been shown that lidar can be manipulated, such that self-driving cars are tricked into taking evasive action.

Biology and Conservation

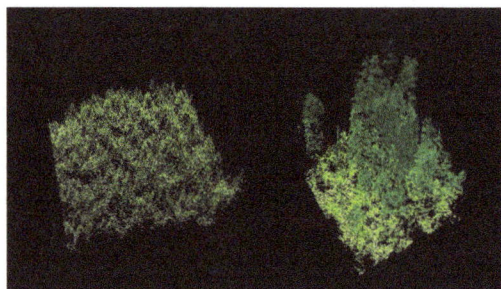

Lidar imaging comparing old-growth forest (right) to a new plantation of trees (left).

Lidar has also found many applications in forestry. Canopy heights, biomass measurements, and leaf area can all be studied using airborne lidar systems. Similarly, lidar is also used by many industries, including Energy and Railroad, and the Department of Transportation as a faster way of surveying. Topographic maps can also be generated readily from lidar, including for recreational use such as in the production of orienteering maps.

In addition, the Save-the-Redwoods League is undertaking a project to map the tall redwoods on the Northern California coast. Lidar allows research scientists to not only measure the height of previously unmapped trees but to determine the biodiversity of the redwood forest. Stephen Sillett, who is working with the League on the North Coast lidar project, claims this technology will be useful in directing future efforts to preserve and protect ancient redwood trees.

Geology and Soil Science

High-resolution digital elevation maps generated by airborne and stationary lidar have led to significant advances in geomorphology (the branch of geoscience concerned with the origin and evolution of the Earth surface topography). The lidar abilities to detect subtle topographic features such as river terraces and river channel banks, to measure the land-surface elevation beneath the vegetation canopy, to better resolve spatial derivatives of elevation, and to detect elevation changes between repeat surveys have enabled many novel studies of the physical and chemical processes that shape landscapes. In 2005 the Tour Ronde in the Mont Blanc massif became the first high alpine mountain on which lidar was employed to monitor the increasing occurrence of severe rock-fall over large rock faces allegedly caused by climate change and degradation of permafrost at high altitude.

In geophysics and tectonics, a combination of aircraft-based lidar and GPS has evolved into an important tool for detecting faults and for measuring uplift. The output of the two technologies can produce extremely accurate elevation models for terrain – models that can even measure ground elevation through trees. This combination was used most famously to find the location of the Seattle Fault in Washington, United States. This combination also measures uplift at Mt. St. Helens by using data from before and after the 2004 uplift. Airborne lidar systems monitor glaciers and have the ability to detect subtle amounts of growth or decline. A satellite-based system, the NASA ICESat, includes a lidar sub-system for this purpose. The NASA Airborne Topographic Mapper is also used extensively to monitor glaciers and perform coastal change analysis. The combination is also used by soil scientists while creating a soil survey. The detailed terrain modeling allows soil scientists to see slope changes and landform breaks which indicate patterns in soil spatial relationships.

Atmospheric Remote Sensing and Meteorology

Initially based on ruby lasers, lidar for meteorological applications was constructed shortly after the invention of the laser and represent one of the first applications of laser technology. Lidar technology has since expanded vastly in capability and lidar systems are used to perform a range of measurements that include profiling clouds, measuring winds, studying aerosols and quantifying various atmospheric components. Atmospheric components can in turn provide useful information including surface pressure (by measuring the absorption of oxygen or nitrogen), greenhouse gas emissions (carbon dioxide and methane), photosynthesis (carbon dioxide), fires (carbon monoxide) and humidity (water vapor). Atmospheric lidars can be either ground-based, airborne or satellite depending on the type of measurement.

Atmospheric lidar remote sensing works in two ways –

1. by measuring backscatter from the atmosphere, and

2. by measuring the scattered reflection off the ground (when the lidar is airborne) or other hard surface.

Backscatter from the atmosphere directly gives a measure of clouds and aerosols. Other derived measurements from backscatter such as winds or cirrus ice crystals require careful selecting of the wavelength and/or polarization detected. *Doppler lidar* and *Rayleigh Doppler lidar* are used to measure temperature and/or wind speed along the beam by measuring the frequency of the backscattered light. The Doppler broadening of gases in motion allows the determination of properties via the resulting frequency shift. Scanning lidars, such as the conical-scanning NASA HARLIE LIDAR, have been used to measure atmospheric wind velocity. The ESA wind mission ADM-Aeolus will be equipped with a Doppler lidar system in order to provide global measurements of vertical wind profiles. A doppler lidar system was used in the 2008 Summer Olympics to measure wind fields during the yacht competition.

Doppler lidar systems are also now beginning to be successfully applied in the renewable energy sector to acquire wind speed, turbulence, wind veer and wind shear data. Both pulsed and continuous wave systems are being used. Pulsed systems use signal timing to obtain vertical distance resolution, whereas continuous wave systems rely on detector focusing.

The term *eolics* has been proposed to describe the collaborative and interdisciplinary study of wind using computational fluid mechanics simulations and Doppler lidar measurements.

The ground reflection of an airborne lidar gives a measure of surface reflectivity (assuming the atmospheric transmittance is well known) at the lidar wavelength. However, the ground reflection is typically used for making absorption measurements of the atmosphere. "Differential absorption lidar" (DIAL) measurements utilize two or more closely spaced (<1 nm) wavelengths to factor out surface reflectivity as well as other transmission losses, since these factors are relatively insensitive to wavelength. When tuned to the appropriate absorption lines of a particular gas, DIAL measurements can be used to determine the concentration (mixing ratio) of that particular gas in the atmosphere. This is referred to as an *Integrated Path Differential Absorption* (IPDA) approach, since it is a measure of the integrated absorption along the entire lidar path. IPDA lidars can be either pulsed or CW and typically use two or more wavelengths. IPDA lidars have been used for remote sensing of carbon dioxide and methane.

Synthetic array lidar allows imaging lidar without the need for an array detector. It can be used for imaging Doppler velocimetry, ultra-fast frame rate (MHz) imaging, as well as for speckle reduction in coherent lidar. An extensive lidar bibliography for atmospheric and hydrospheric applications is given by Grant.

Scheimpflug Principles

Another lidar technique for atmospheric remote sensing has emerged. It is based on Scheimpflug principles referred to as scheimpflug lidar (slidar).

"The implication of the Scheimpflug principle is that when a laser beam is transmitted into the atmosphere, the backscattering echo of the entire illuminating probe volume is still in focus simultaneously without diminishing the aperture as long as the object plane, image plane and the lens plane intersect with each other". A two dimensional CCD/CMOS camera is used to resolve the backscattering echo of the transmitted laser beam.

Thus as in the case of conventional lidar technologies continuous wave light sources such as diode lasers can be employed for remote sensing instead of using complicated nano second pulse light sources. The SLidar system is also a robust and inexpensive system based on compact laser diodes and array detectors.

Law Enforcement

Lidar speed guns are used by the police to measure the speed of vehicles for speed limit enforcement purposes.

Military

Few military applications are known to be in place and are classified (like the lidar-based speed measurement of the AGM-129 ACM stealth nuclear cruise missile), but a considerable amount of research is underway in their use for imaging. Higher resolution systems collect enough detail to identify targets, such as tanks. Examples of military applications of lidar include the Airborne Laser Mine Detection System (ALMDS) for counter-mine warfare by Areté Associates.

A NATO report (RTO-TR-SET-098) evaluated the potential technologies to do stand-off detection for the discrimination of biological warfare agents. The potential technologies evaluated were Long-Wave Infrared (LWIR), Differential Scattering (DISC), and Ultraviolet Laser Induced Fluorescence (UV-LIF). The report concluded that : *Based upon the results of the lidar systems tested and discussed above, the Task Group recommends that the best option for the near-term (2008–2010) application of stand-off detection systems is UV LIF* . However, in the long-term, other techniques such as stand-off Raman spectroscopy may prove to be useful for identification of biological warfare agents.

Short-range compact spectrometric lidar based on Laser-Induced Fluorescence (LIF) would address the presence of bio-threats in aerosol form over critical indoor, semi-enclosed and outdoor venues like stadiums, subways, and airports. This near real-time capability would enable rapid detection of a bioaerosol release and allow for timely implementation of measures to protect occupants and minimize the extent of contamination.

The Long-Range Biological Standoff Detection System (LR-BSDS) was developed for the US Army to provide the earliest possible standoff warning of a biological attack. It is an airborne system carried by a helicopter to detect man-made aerosol clouds containing biological and chemical agents at long range. The LR-BSDS, with a detection range of 30 km or more, was fielded in June 1997. Five lidar units produced by the German company Sick AG were used for short range detection on Stanley, the autonomous car that won the 2005 DARPA Grand Challenge.

A robotic Boeing AH-6 performed a fully autonomous flight in June 2010, including avoiding obstacles using lidar.

Mining

For The calculation of ore volumes is accomplished by periodic (monthly) scanning in areas of ore removal, then comparing surface data to the previous scan.

Lidar sensors may also be used for obstacle detection and avoidance for robotic mining vehicles such as in the Komatsu Autonomous Haulage System (AHS) used in Rio Tinto's Mine of the Future.

Physics and Astronomy

A worldwide network of observatories uses lidars to measure the distance to reflectors placed on the moon, allowing the position of the moon to be measured with mm precision and tests of general relativity to be done. MOLA, the Mars Orbiting Laser Altimeter, used a lidar instrument in a Mars-orbiting satellite (the NASA Mars Global Surveyor) to produce a spectacularly precise global topographic survey of the red planet.

In September, 2008, the NASA Phoenix Lander used lidar to detect snow in the atmosphere of Mars.

In atmospheric physics, lidar is used as a remote detection instrument to measure densities of certain constituents of the middle and upper atmosphere, such as potassium, sodium, or molecular nitrogen and oxygen. These measurements can be used to calculate temperatures. Lidar can also be used to measure wind speed and to provide information about vertical distribution of the aerosol particles.

At the JET nuclear fusion research facility, in the UK near Abingdon, Oxfordshire, lidar Thomson Scattering is used to determine Electron Density and Temperature profiles of the plasma.

Rock Mechanics

Lidar has been widely used in rock mechanics for rock mass characterization and slope change detection. Some important geomechanical properties from the rock mass can be extracted from the 3D point clouds obtained by means of the lidar. Some of these properties are:

- Discontinuity orientation
- Discontinuity spacing and RQD
- Discontinuity aperture
- Discontinuity persistence
- Discontinuity roughness
- Water infiltration

Some of these properties have been used to assess the geomechanical quality of the rock mass through the RMR index. Moreover, as the orientations of discontinuities can be extracted using the existing methodologies, it is possible to assess the geomechanical quality of a rock slope through the SMR index. In addition to this, the comparison of different 3D point clouds from a slope acquired at different times allows to study the changes produced on the scene during this time interval as a result of rockfalls or any other landsliding processes.

THOR

THOR is a laser designed towards measuring Earth's atmospheric conditions. The laser enters a cloud cover and measures the thickness of the return halo. The sensor has a fiber optic aperture with a width of 7.5 inches that is used to measure the return light.

Robotics

Lidar technology is being used in robotics for the perception of the environment as well as object classification. The ability of lidar technology to provide three-dimensional elevation maps of the terrain, high precision distance to the ground, and approach velocity can enable safe landing of robotic and manned vehicles with a high degree of precision.

Spaceflight

Lidar is increasingly being utilized for rangefinding and orbital element calculation of relative velocity in proximity operations and stationkeeping of spacecraft. Lidar has also been used for atmospheric studies from space. Short pulses of laser light beamed from a spacecraft can reflect off of tiny particles in the atmosphere and back to a telescope aligned with the spacecraft laser. By precisely timing the lidar 'echo,' and by measuring how much laser light is received by the telescope, scientists can accurately determine the location, distribution and nature of the particles. The result is a revolutionary new tool for studying constituents in the atmosphere, from cloud droplets to industrial pollutants, that are difficult to detect by other means."

Surveying

This TomTom mapping van is fitted with five lidars on its roof rack.

Airborne lidar sensors are used by companies in the remote sensing field. They can be used to create a DTM (Digital Terrain Model) or DEM (Digital Elevation Model); this is quite a common practice for larger areas as a plane can acquire 3–4 km wide swaths in a single flyover. Greater vertical accuracy of below 50 mm can be achieved with a lower flyover, even in forests, where it is able to give the height of the canopy as well as the ground elevation. Typically, a GNSS receiver configured over a georeferenced control point is needed to link the data in with the WGS (World Geodetic System).

Forestry

Lidar systems have also been applied to improve forestry management. Measurements are used to take inventory in forest plots as well as calculate individual tree heights, crown width and crown diameter. Other statistical analysis use lidar data to estimate total plot information like canopy volume, mean, minimum and maximum heights, and vegetation cover estimates.

Transport

Lidar has been used in the railroad industry to generate asset health reports for asset management and by departments of transportation to assess their road conditions. CivilMaps.com is a leading company in the field. Lidar has been used in adaptive cruise control (ACC) systems for automobiles. Systems such as those by Siemens, Hella and Cepton use a lidar device mounted on the front of the vehicle, such as the bumper, to monitor the distance between the vehicle and any vehicle in front of it. In the event the vehicle in front slows down or is too close, the ACC applies the brakes to slow the vehicle. When the road ahead is clear, the ACC allows the vehicle to accelerate to a speed preset by the driver. A lidar-based device, the Ceilometer is used at airports worldwide to measure the height of clouds on runway approach paths.

Wind Farm Optimization

Lidar can be used to increase the energy output from wind farms by accurately measuring wind speeds and wind turbulence. Experimental lidar systems can be mounted on the nacelle of a wind turbine or integrated into the rotating spinner to measure oncoming horizontal winds, winds in the wake of the wind turbine, and proactively adjust blades to protect components and increase power. Lidar is also used to characterise the incident wind resource for comparison with wind turbine power production to verify the performance of the wind turbine by measuring the wind turbine's power curve. Wind farm optimization can be considered a topic in *applied eolics*.

Solar Photovoltaic Deployment Optimization

Lidar can also be used to assist planners and developers in optimizing solar photovoltaic systems at the city level by determining appropriate roof tops and for determining shading losses. Recent airborn laser scanning efforts have focused on ways to estimate the amount of solar light hitting vertical building facades, or by incorporating more detailed shading losses by considering the influence from vegetation and larger surrounding terrain.

Video Games

Racing game iRacing features scanned tracks, resulting in bumps with millimeter precision in the in-game 3D mapping environment.

The 2017 exploration game *Scanner Sombre*, by Introversion Software, uses Lidar as a fundamental game mechanic.

Other Uses

The video for the song "House of Cards" by Radiohead was believed to be the first use of real-time 3D laser scanning to record a music video. The range data in the video is not completely from a lidar, as structured light scanning is also used.

Alternative Technologies

Recent development of Structure From Motion (SFM) technologies allows delivering 3D images and maps based on data extracted from visual and IR photography. The elevation or 3D data is extracted using multiple parallel passes over mapped area, yielding both visual light images and 3D structure from the same sensor, which is often a specially chosen and calibrated digital camera.

Interferometric Synthetic Aperture Radar

Interferometric synthetic aperture radar, abbreviated InSAR (or deprecated IfSAR), is a radar technique used in geodesy and remote sensing. This geodetic method uses two or more synthetic aperture radar (SAR) images to generate maps of surface deformation or digital elevation, using differences in the phase of the waves returning to the satellite or aircraft. The technique can potentially measure millimetre-scale changes in deformation over spans of days to years. It has applications for geophysical monitoring of natural hazards, for example earthquakes, volcanoes and landslides, and in structural engineering, in particular monitoring of subsidence and structural stability.

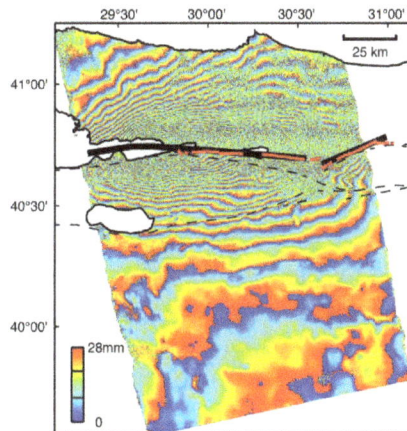

Interferogram produced using ERS-2 data from 13 August and 17 September 1999, spanning the 17 August Izmit (Turkey) earthquake. (NASA/JPL-Caltech)

Technique

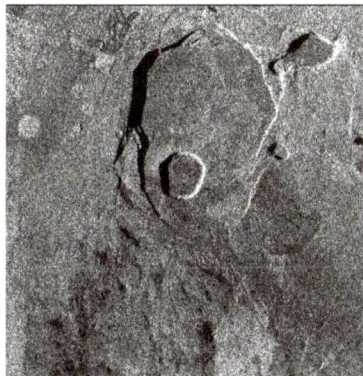

SAR amplitude image of Kīlauea (NASA/JPL-Caltech)

Synthetic Aperture Radar

Synthetic aperture radar (SAR) is a form of radar in which sophisticated processing of radar data is used to produce a very narrow effective beam. It can be used to form images of relatively immobile targets; moving targets can be blurred or displaced in the formed images. SAR is a form of active remote sensing – the antenna transmits radiation that is reflected from the image area, as opposed to passive sensing, where the reflection is detected from ambient illumination. SAR image acquisition is therefore independent of natural illumination and images can be taken at night. Radar uses electromagnetic radiation at microwave frequencies; the atmospheric absorption at typical radar wavelengths is very low, meaning observations are not prevented by cloud cover.

Phase

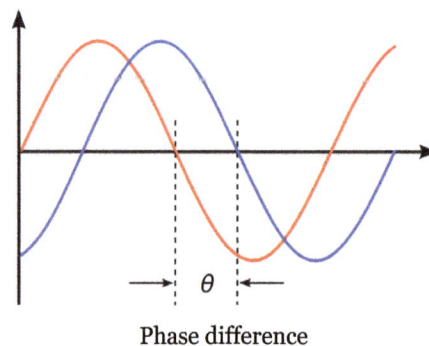

Phase difference

SAR makes use of the amplitude and the absolute phase of the return signal data. In contrast, interferometry uses differential phase of the reflected radiation, either from multiple passes along the same trajectory and/or from multiple displaced phase centers (antennas) on a single pass. Since the outgoing wave is produced by the satellite, the phase is known, and can be compared to the phase of the return signal. The phase of the return wave depends on the distance to the ground, since the path length to the ground and back will consist of a number of whole wavelengths plus some fraction of a wavelength. This is observable as a phase difference or phase shift in the returning wave. The total distance to the satellite (i.e., the number of whole wavelengths) is known based on the time that it takes for the energy to make the round trip back to the satellite—but it is the extra fraction of a wavelength that is of particular interest and is measured to great accuracy.

In practice, the phase of the return signal is affected by several factors, which together can make the absolute phase return in any SAR data collection essentially arbitrary, with no correlation from pixel to pixel. To get any useful information from the phase, some of these effects must be isolated and removed. Interferometry uses two images of the same area taken from the same position (or, for topographic applications, slightly different positions) and finds the difference in phase between them, producing an image known as an interferogram. This is measured in radians of phase difference and, because of the cyclic nature of phase, is recorded as repeating fringes that each represent a full 2π cycle.

Factors Affecting Phase

The most important factor affecting the phase is the interaction with the ground surface. The phase of the wave may change on reflection, depending on the properties of the material. The reflected signal back from any one pixel is the summed contribution to the phase from many smaller 'targets' in that ground area, each with different dielectric properties and distances from the satellite, meaning the returned signal is arbitrary and completely uncorrelated with that from adjacent pixels. Importantly though, it is consistent – provided nothing on the ground changes the contributions from each target should sum identically each time, and hence be removed from the interferogram.

Once the ground effects have been removed, the major signal present in the interferogram is a contribution from orbital effects. For interferometry to work, the satellites must be as close as possible to the same spatial position when the images are acquired. This means that images from two satellite platforms with different orbits cannot be compared, and for a given satellite data from the same orbital track must be used. In practice the perpendicular distance between them, known as the *baseline*, is often known to within a few centimetres but can only be controlled on a scale of tens to hundreds of metres. This slight difference causes a regular difference in phase that changes smoothly across the interferogram and can be modelled and removed.

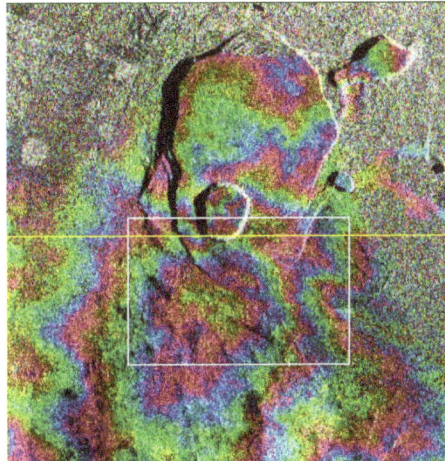

Corresponding interferogram of Kīlauea, showing topographic fringes (NASA/JPL-Caltech)

The slight difference in satellite position also alters the distortion caused by topography, meaning an extra phase difference is introduced by a stereoscopic effect. The longer the baseline, the smaller the topographic height needed to produce a fringe of phase change – known as the *altitude of ambiguity*. This effect can be exploited to calculate the topographic height, and used to produce a digital elevation model (DEM).

If the height of the topography is already known, the topographic phase contribution can be calculated and removed. This has traditionally been done in two ways. In the *two-pass* method, elevation data from an externally derived DEM is used in conjunction with the orbital information to calculate the phase contribution. In the *three-pass* method two images acquired a short time apart are used to create an interferogram, which is assumed to have no deformation signal and therefore represent the topographic contribution. This interferogram is then subtracted from a third image with a longer time separation to give the residual phase due to deformation.

Once the ground, orbital and topographic contributions have been removed the interferogram contains the deformation signal, along with any remaining noise. The signal measured in the interferogram represents the change in phase caused by an increase or decrease in distance from the ground pixel to the satellite, therefore only the component of the ground motion parallel to the satellite line of sight vector will cause a phase difference to be observed. For sensors like ERS with a small incidence angle this measures vertical motion well, but is insensitive to horizontal motion perpendicular to the line of sight (approximately north-south). It also means that vertical motion and components of horizontal motion parallel to the plane of the line of sight (approximately east-west) cannot be separately resolved.

One fringe of phase difference is generated by a ground motion of half the radar wavelength, since this corresponds to a whole wavelength increase in the two-way travel distance. Phase shifts are only resolvable relative to other points in the interferogram. Absolute deformation can be inferred by assuming one area in the interferogram (for example a point away from expected deformation sources) experienced no deformation, or by using a ground control (GPS or similar) to establish the absolute movement of a point.

Difficulties

A variety of factors govern the choice of images which can be used for interferometry. The simplest is data availability – radar instruments used for interferometry commonly don't operate continuously, acquiring data only when programmed to do so. For future requirements it may be possible to request acquisition of data, but for many areas of the world archived data may be sparse. Data availability is further constrained by baseline criteria. Availability of a suitable DEM may also be a factor for two-pass InSAR; commonly 90 m SRTM data may be available for many areas, but at high latitudes or in areas of poor coverage alternative datasets must be found.

A fundamental requirement of the removal of the ground signal is that the sum of phase contributions from the individual targets within the pixel remains constant between the two images and is completely removed. However, there are several factors that can cause this criterion to fail. Firstly the two images must be accurately co-registered to a sub-pixel level to ensure that the same ground targets are contributing to that pixel. There is also a geometric constraint on the maximum length of the baseline – the difference in viewing angles must not cause phase to change over the width of one pixel by more than a wavelength. The effects of topography also influence the condition, and baselines need to be shorter if terrain gradients are high. Where co-registration is poor or the maximum baseline is exceeded the pixel phase will become incoherent – the phase becomes essentially random from pixel to pixel rather than varying smoothly, and the area appears noisy. This is also true for anything else that changes the contributions to the phase within each pixel, for example changes to the ground targets in each pixel caused by vegetation growth, landslides, agriculture or snow cover.

Another source of error present in most interferograms is caused by the propagation of the waves through the atmosphere. If the wave travelled through a vacuum it should theoretically be possible (subject to sufficient accuracy of timing) to use the two-way travel-time of the wave in combination with the phase to calculate the exact distance to the ground. However, the velocity of the wave through the atmosphere is lower than the speed of light in a vacuum, and depends on air temperature, pressure and the partial pressure of water vapour. It is this unknown phase delay that prevents the integer number of wavelengths being calculated. If the atmosphere was horizontally homogeneous over the length scale of an interferogram and vertically over that of the topography then the effect would simply be a constant phase difference between the two images which, since phase difference is measured relative to other points in the interferogram, would not contribute to the signal. However, the atmosphere is laterally heterogeneous on length scales both larger and smaller than typical deformation signals. This spurious signal can appear completely unrelated to the surface features of the image, however, in other cases the atmospheric phase delay is caused by vertical inhomogeneity at low altitudes and this may result in fringes appearing to correspond with the topography.

Persistent Scatterer InSAR

Persistent or Permanent Scatterer techniques are a relatively recent development from conventional InSAR, and rely on studying pixels which remain coherent over a sequence of interferograms. In 1999, researchers at Politecnico di Milano, Italy, developed a new multi-image approach in which one searches the stack of images for objects on the ground providing consistent and stable radar reflections back to the satellite. These objects could be the size of a pixel or, more commonly, sub-pixel sized, and are present in every image in the stack. That specific implementation is patented.

Some research centres and companies, were inspired to develop variations of their own algorithms which would also overcome InSAR's limitations. In scientific literature, these techniques are collectively referred to as Persistent Scatterer Interferometry or PSI techniques. The term Persistent Scatterer Interferometry (PSI) was proposed by European Space Agency (ESA) to define the second generation of radar interferometry techniques. This term is nowadays commonly accepted by scientific and the end user community.

Commonly such techniques are most useful in urban areas with lots of permanent structures, for example the PSI studies of European geohazard sites undertaken by the Terrafirma project. The Terrafirma project provides a ground motion hazard information service, distributed throughout Europe via national geological surveys and institutions. The objective of this service is to help save lives, improve safety, and reduce economic loss through the use of state-of-the-art PSI information. Over the last 9 years this service has supplied information relating to urban subsidence and uplift, slope stability and landslides, seismic and volcanic deformation, coastlines and flood plains.

Producing Interferograms

The processing chain used to produce interferograms varies according to the software used and the precise application but will usually include some combination of the following steps.

Two SAR images are required to produce an interferogram; these may be obtained pre-processed, or produced from raw data by the user prior to InSAR processing. The two images must first be co-registered, using a correlation procedure to find the offset and difference in geometry between the two amplitude images. One SAR image is then re-sampled to match the geometry of the other, meaning each pixel represents the same ground area in both images. The interferogram is then formed by cross-multiplication of each pixel in the two images, and the interferometric phase due to the curvature of the Earth is removed, a process referred to as flattening. For deformation applications a DEM can be used in conjunction with the baseline data to simulate the contribution of the topography to the interferometric phase, this can then be removed from the interferogram.

Once the basic interferogram has been produced, it is commonly filtered using an adaptive power-spectrum filter to amplify the phase signal. For most quantitative applications the consecutive fringes present in the interferogram will then have to be *unwrapped*, which involves interpolating over the 0 to 2π phase jumps to produce a continuous deformation field. At some point, before or after unwrapping, incoherent areas of the image may be masked out. The final processing stage involves geocoding the image, which resamples the interferogram from the acquisition geometry (related to direction of satellite path) into the desired geographic projection.

Hardware

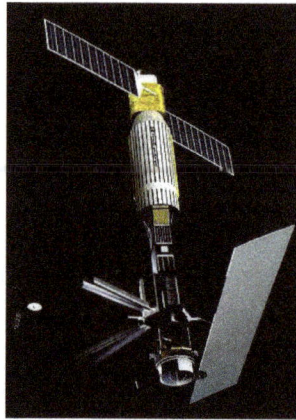

Seasat (NASA/JPL-Caltech)

Spaceborne

Early exploitation of satellite-based InSAR included use of Seasat data in the 1980s, but the potential of the technique was expanded in the 1990s, with the launch of ERS-1 (1991), JERS-1 (1992), RADARSAT-1 and ERS-2 (1995). These platforms provided the stable, well-defined orbits and short baselines necessary for InSAR. More recently, the 11-day NASA STS-99 mission in February 2000 used a SAR antenna mounted on the space shuttle to gather data for the Shuttle Radar Topography Mission. In 2002 ESA launched the ASAR instrument, designed as a successor to ERS, aboard Envisat. While the majority of InSAR to date has utilised the C-band sensors, recent missions such as the ALOS PALSAR, TerraSAR-X and COSMO-SkyMed are expanding the available data in the L- and X-band.

Most recently, ESA launched Sentinel-1A and Sentinel-1B – two C-band sensors. Together, they provide InSAR coverage on a global scale and on a 6-day repeat cycle.

Airborne

Airborne InSAR data acquisition systems are built by companies such as the American Intermap, the German AeroSensing, and the Brazilian OrbiSat.

Terrestrial or Ground-based

Terrestrial or ground-based SAR Interferometry (GBInSAR or TInSAR) is a remote sensing technique for the displacement monitoring of slopes, rock scarps, volcanoes, landslides, buildings, infrastructures etc. This technique is based on the same operational principles of the Satellite SAR Interferometry, but the Synthetic Aperture of the Radar (SAR) is obtained by an antenna moving on a rail instead of a satellite moving around an orbit. SAR technique allows 2D radar image of the investigated scenario to be achieved, with a high range resolution (along the instrumental line of sight) and cross-range resolution (along the scan direction). Two antennas respectively emit and receive microwave signals and, by calculating the phase difference between two measurements taken in two different times, it is possible to compute the displacement of all the pixels of the SAR image. The accuracy in the displacement measurement is of the same order of magnitude as the EM wavelength and depends also on the specific local and atmospheric conditions.

A deformation plot showing slope instability using Terrestrial InSAR

Applications

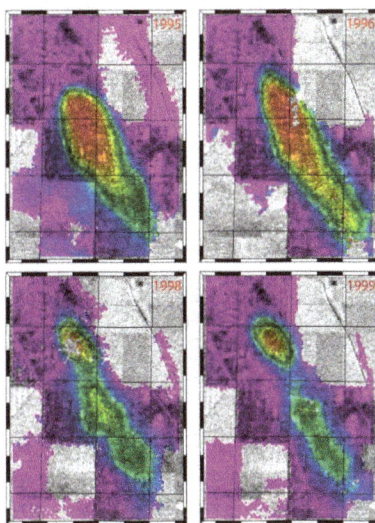

Rapid ground subsidence over the Lost Hills oil field in California. (NASA/JPL-Caltech)

Tectonic

InSAR can be used to measure tectonic deformation, for example ground movements due to earthquakes. It was first used for the 1992 Landers earthquake, but has since been utilised extensively for a wide variety of earthquakes all over the world. In particular the 1999 Izmit and 2003 Bam earthquakes were extensively studied. InSAR can also be used to monitor creep and strain accumulation on faults.

Volcanic

InSAR can be used in a variety of volcanic settings, including deformation associated with eruptions, inter-eruption strain caused by changes in magma distribution at depth, gravitational spreading of volcanic edifices, and volcano-tectonic deformation signals. Early work on volcanic InSAR included studies on Mount Etna, and Kilauea, with many more volcanoes being studied as the field developed. The technique is now widely used for academic research into volcanic deformation, although its use as an operational monitoring technique for volcano observatories has been limited by issues such as orbital repeat times, lack of archived data, coherence and atmospheric errors. Recently InSAR has been used to study rifting processes in Ethiopia.

Subsidence

Ground subsidence from a variety of causes has been successfully measured using InSAR, in particular subsidence caused by oil or water extraction from underground reservoirs, subsurface mining and collapse of old mines. Thus, InSAR has become an indispensable tool to satisfactorily address many subsidence studies. Tomás et al. performed a cost analysis that allowed to identify the strongest points of InSAR techniques compared with other conventional techniques: (1) higher data acquisition frequency and spatial coverage; and (2) lower annual cost per measurement point and per square kilometre.

Landslides

Although InSAR technique can present some limitations when applied to landslides, it can also be used for monitoring landscape features such as landslides.

Ice Flow

Glacial motion and deformation have been successfully measured using satellite interferometry. The technique allows remote, high-resolution measurement of changes in glacial structure, ice flow, and shifts in ice dynamics, all of which agree closely with ground observations.

Kamchatka Peninsula, Landsat data draped over SRTM digital elevation model (NASA/JPL-Caltech)

Infrastructure and Building Monitoring

InSAR can also be used to monitor the stability of built structures,. Especially Very High Resolution SAR data (such as derived from the TerraSAR-X StripMap mode or COSMO-Skymed HIMAGE mode) are suitable for this task. InSAR is used for monitoring highway and railway settlements, dike stability, forensic engineering and many other uses.

DEM Generation

Interferograms can be used to produce digital elevation maps (DEMs) using the stereoscopic effect caused by slight differences in observation position between the two images. When using two images produced by the same sensor with a separation in time, it must be assumed other phase contributions (for example from deformation or atmospheric effects) are minimal. In 1995 the two ERS satellites flew in tandem with a one-day separation for this purpose. A second approach is to use two antennas mounted some distance apart on the same platform, and acquire the images at the same time, which ensures no atmospheric or deformation signals are present. This approach was followed by NASA's SRTM mission aboard the space shuttle in 2000. InSAR-derived DEMs can be used for later two-pass deformation studies, or for use in other geophysical applications.

Geodetic Control Network

A geodetic control network (also geodetic network, reference network, control point network, or control network) is a network, of often of triangles, which are measured exactly by techniques of terrestrial surveying or by satellite geodesy.

Control point marker placed by the US Coast and Geodetic Survey	Example of triangle network and its application in chartography	Typical GNSS reference station
Worldwide BC-4 camera geometric satellite triangulation network	International Terrestrial Reference System (ITRF) reference stations	Network of reference stations used by Austrian Positioning Service (APOS)

A geodetic control network consists of stable, identifiable points with published datum values derived from observations that tie the points together.

Classically, a control is divided into horizontal (X-Y) and vertical (Z) controls (components of the control), however with the advent of satellite navigation systems, GPS in particular, this division is becoming obsolete.

Many organizations contribute information to the geodetic control network.

The higher-order (high precision, usually millimeter-to-decimeter on a scale of continents) control points are normally defined in both space and time using global or space techniques, and are used for "lower-order" points to be tied into. The lower-order control points are normally used for engineering, construction and navigation. The scientific discipline that deals with the establishing of coordinates of points in a high-order control network is called geodesy, and the technical discipline that does the same for points in a low-order control network is called surveying.

Chartography

After a cartographer registers key points in a digital map to the real world coordinates of those points on the ground, the map is then said to be "in control". Having a base map and other data in geodetic control means that they will overlay correctly.

When map layers are not in control, it requires extra work to adjust them to line up, which introduces additional error. Those real world coordinates are generally in some particular map projection, unit, and geodetic datum.

Triangulation

In "classical geodesy" (up to the sixties) control networks was established by triangulation using measurements of angles and of some spare distances. The precise orientation to the geographic north is achieved through methods of geodetic astronomy. The principal instruments used are theodolites and tacheometers, which nowadays are equipped with infrared distance measuring, data bases, communication systems and partly by satellite links.

Trilateration

Electronic distance measurement (EDM) was introduced around 1960, when the prototype instruments became small enough to be used in the field. Instead of using only sparse and much less accurate distance measurements some control networks was established or updated by using trilateration more accurate distance measurements than was previously possible and no angle measurements.

EDM increased network accuracies up to 1:1 million (1 cm per 10 km; today at least 10 times better), and made surveying less costly.

Satellite Geodesy

The geodetic use of satellites began around the same time. By using bright satellites like Echo I, Echo II and Pageos, global networks were determined, which later provided support for the theory of plate tectonics.

Another important improvement was the introduction of radio and electronic satellites like Geos A and B (1965–70), of the Transit system (Doppler effect) 1967-1990 — which was the predecessor of GPS - and of laser techniques like Lageos (USA) or Starlette (F). Despite the use of spacecraft, small networks for cadastral and technical projects are mainly measured terrestrially, but in many cases incorporated in national and global networks by satellite geodesy.

Global Navigation Satellite Systems (GNSS)

Nowadays, several hundred geodetic satellites are in orbit, supplemented by a large number of remote sensing satellites and navigation systems like GPS and Glonass, which will be followed by the European Galileo satellites in 2013.

While these developments have made satellite-based geodetic network surveying more flexible and cost effective than its terrestrial equivalent, the continued existence of fixed point networks is still needed for administrative and legal purposes on local and regional scales. Global geodetic networks cannot be defined to be fixed, since geodynamics are continuously changing the position of all continents by 2 to 20 cm per year. Therefore, modern global networks like ETRS89 or ITRF show not only coordinates of their "fixed points", but also their annual velocities.

References

- Harvey, Brian (2007). "Military programs". The Rebirth of the Russian Space Program (1st ed.). Germany: Springer. ISBN 978-0-387-71354-0

- Tang, Lina; Shao, Guofan (2015-06-21). "Drone remote sensing for forestry research and practices". Journal of Forestry Research. 26 (4): 791–797. ISSN 1007-662X. doi:10.1007/s11676-015-0088-y

- Pike, John. "BeiDou (Big Dipper)". Space. GlobalSecurity.org. Archived from the original on 28 November 2006. Retrieved 2006-11-09

- GNSS Positioning Approaches – GPS Satellite Surveying, Fourth Edition – Leick. Wiley Online Library. pp. 257–399. doi:10.1002/9781119018612.ch6

- O'Leary, Beth Laura; Darrin, Ann Garrison (2009). Handbook of Space Engineering, Archaeology, and Heritage. Hoboken: CRC Press. pp. 239–240. ISBN 9781420084320

- Yu, B.; Liu, G.; Zhang, R.; Jia, H.; Li, T.; Wang, X.; Dai, K.; Ma, D. (2013). "Monitoring subsidence rates along road network by persistent scatterer SAR interferometry with high-resolution TerraSAR-X imagery". Journal of Modern Transportation. 21: 236–246. doi:10.1007/s40534-013-0030-y

- "US Department of Defense Reports on China's Space Capabilities". Space Safety Magazine. 27 May 2013. Retrieved 26 June 2013

- Medina A, Gayá F, Pozo F (2006). "Compact laser radar and three-dimensional camera". J. Opt. Soc. Am. A. 23: 800–805. Bibcode:2006JOSAA..23..800M. doi:10.1364/josaa.23.000800

- Michael Russell Rip; James M. Hasik (2002). The Precision Revolution: GPS and the Future of Aerial Warfare. Naval Institute Press. p. 65. ISBN 1-55750-973-5. Retrieved January 14, 2010

- "Rediscovering the lost archaeological landscape of southern New England using airborne light detection and ranging (LiDAR)". Journal of Archaeological Science. 43: 9–20. doi:10.1016/j.jas.2013.12.004. Retrieved 2016-02-22

- "China sends Beidou navigation satellite to orbit". Spaceflight Now. 2010-06-02. Archived from the original on 5 June 2010. Retrieved 2010-06-04

- Evans, D.H.; Fletcher, R.J.; et al. "Uncovering archaeological landscapes at Angkor using lidar". PNAS. 110 (31): 12595–12600. PMC 3732978. PMID 23847206. doi:10.1073/pnas.1306539110

- Michael Russell Rip; James M. Hasik (2002). The Precision Revolution: GPS and the Future of Aerial Warfare. Naval Institute Press. ISBN 1-55750-973-5. Retrieved May 25, 2008

- "BeiDou-1 commercial controversy: 10 times the price of GPS terminal" (in Chinese). NetEase. 2008-06-28. Retrieved 2010-05-23

- Riquelme, A.J.; Abellán, A.; Tomás, R. (2015). "Discontinuity spacing analysis in rock masses using 3D point clouds". Engineering Geology. 195: 185–195. doi:10.1016/j.enggeo.2015.06.009

5

Essential Aspects of Geodesy

The exact calculation of various locations on Earth require the study of the shape of the Earth. The Earth's figure is theoretically explained as an oblate ellipsoid due to its rotation. Geodesy is best understood in confluence with the major topics listed in the following chapter.

Geodesics on an Ellipsoid

The study of geodesics on an ellipsoid arose in connection with geodesy specifically with the solution of triangulation networks. The figure of the Earth is well approximated by an *oblate ellipsoid*, a slightly flattened sphere. A *geodesic* is the shortest path between two points on a curved surface, i.e., the analogue of a straight line on a plane surface. The solution of a triangulation network on an ellipsoid is therefore a set of exercises in spheroidal trigonometry (Euler 1755).

If the Earth is treated as a sphere, the geodesics are great circles (all of which are closed) and the problems reduce to ones in spherical trigonometry. However, Newton (1687) showed that the effect of the rotation of the Earth results in its resembling a slightly oblate ellipsoid and, in this case, the equator and the meridians are the only closed geodesics. Furthermore, the shortest path between two points on the equator does not necessarily run along the equator. Finally, if the ellipsoid is further perturbed to become a triaxial ellipsoid (with three distinct semi-axes), only three geodesics are closed.

Geodesics on an Ellipsoid of Revolution

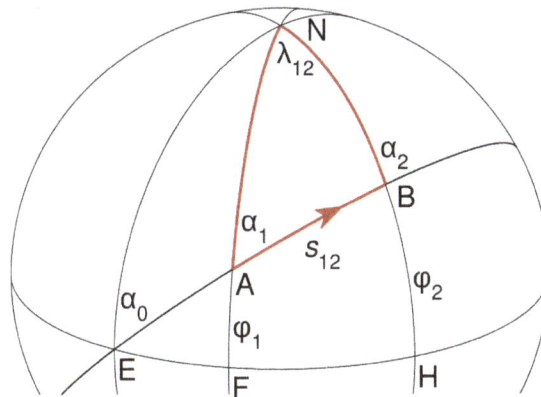

Fig. a. A geodesic *AB* on an ellipsoid of revolution. *N* is the north pole and *EFH* lie on the equator.

There are several ways of defining geodesics (Hilbert & Cohn-Vossen 1952). A simple definition is as the shortest path between two points on a surface. However, it is frequently more useful to define them as paths with zero geodesic curvature—i.e., the analogue of straight lines on a curved surface.

This definition encompasses geodesics traveling so far across the ellipsoid's surface (somewhat more than half the circumference) that other distinct routes require less distance. Locally, these geodesics are still identical to the shortest distance between two points.

By the end of the 18th century, an ellipsoid of revolution (the term spheroid is also used) was a well-accepted approximation to the figure of the Earth. The adjustment of triangulation networks entailed reducing all the measurements to a reference ellipsoid and solving the resulting two-dimensional problem as an exercise in spheroidal trigonometry (Bomford 1952) (Leick et al. 2015).

It is possible to reduce the various geodesic problems into one of two types. Consider two points: A at latitude φ_1 and longitude λ_1 and B at latitude φ_2 and longitude λ_2. The connecting geodesic (from A to B) is AB, of length s_{12}, which has azimuths α_1 and α_2 at the two endpoints. The two geodesic problems usually considered are:

1. the *direct geodesic problem* or *first geodesic problem*, given A, α_1, and s_{12}, determine B and α_2;

2. the *inverse geodesic problem* or *second geodesic problem*, given A and B, determine s_{12}, α_1, and α_2.

As can be seen from Fig. a, these problems involve solving the triangle NAB given one angle, α_1 for the direct problem and $\lambda_{12} = \lambda_2 - \lambda_1$ for the inverse problem, and its two adjacent sides. For a sphere the solutions to these problems are simple exercises in spherical trigonometry, whose solution is given by formulas for solving a spherical triangle.

For an ellipsoid of revolution, the characteristic constant defining the geodesic was found by Clairaut (1735). A systematic solution for the paths of geodesics was given by Legendre (1806) and Oriani (1806) (and subsequent papers in 1808 and 1810). The full solution for the direct problem (complete with computational tables and a worked out example) is given by Bessel (1825).

During the 18th century geodesics were typically referred to as "shortest lines". The term "geodesic line" was coined by Laplace (1799b):

Nous désignerons cette ligne sous le nom de *ligne géodésique* [We will call this line the *geodesic line*].

This terminology was introduced into English either as "geodesic line" or as "geodetic line", for example (Hutton 1811),

A line traced in the manner we have now been describing, or deduced from trigonometrical measures, by the means we have indicated, is called a *geodetic* or *geodesic line:* it has the property of being the shortest which can be drawn between its two extremities on the surface of the Earth; and it is therefore the proper itinerary measure of the distance between those two points.

In its adoption by other fields "geodesic line", frequently shortened, to "geodesic", was preferred.

Equations for a Geodesic

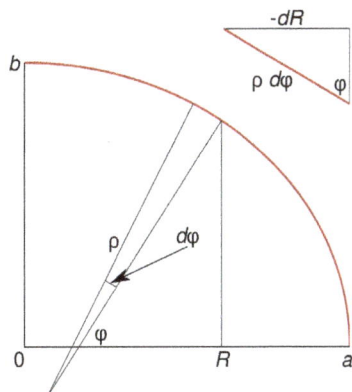

Fig. b. Differential element of a meridian ellipse.

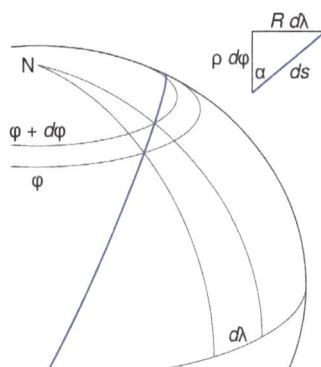

Fig. c. Differential element of a geodesic on an ellipsoid.

Here the equations for a geodesic are developed; the derivation closely follows that of Bessel (1825). Jordan & Eggert (1941), Bagratuni (1962), Gan'shin (1967), Krakiwsky & Thomson (1974), Rapp (1993), Jekeli (2012), and Borre & Strang (2012) also provide derivations of these equations.

Consider an ellipsoid of revolution with equatorial radius a and polar semi-axis b. Define the flattening $f = (a - b)/a$, the eccentricity $e = \sqrt{a^2 - b^2} / a = \sqrt{f(2 - f)}$, and the second eccentricity $e' = \sqrt{a^2 - b^2} / b = e/(1 - f)$. (In most applications in geodesy, the ellipsoid is taken to be oblate, $a > b$; however, the theory applies without change to prolate ellipsoids, $a < b$, in which case f, e^2, and e'^2 are negative.)

Let an elementary segment of a path on the ellipsoid have length ds. From Figs. b and c, we see that if its azimuth is α, then ds is related to $d\varphi$ and $d\lambda$ by

$$(1) \qquad \cos\alpha\, ds = \rho d\varphi = -dR / \sin\varphi, \quad \sin\alpha\, ds = R d\lambda,$$

where ρ is the meridional radius of curvature, $R = \nu \cos\varphi$ is the radius of the circle of latitude φ, and ν is the normal radius of curvature. The elementary segment is therefore given by

$$ds^2 = \rho^2 d\varphi^2 + R^2 d\lambda^2$$

or

$$ds = \sqrt{\rho^2 \varphi'^2 + R^2}\, d\lambda$$
$$\equiv L(\varphi, \varphi')d\lambda,$$

where $\varphi' = d\varphi/d\lambda$ and the Lagrangian function L depends on φ through $\rho(\varphi)$ and $R(\varphi)$. The length of an arbitrary path between (φ_1, λ_1) and (φ_2, λ_2) is given by

$$s_{12} = \int_{\lambda_1}^{\lambda_2} L(\varphi, \varphi')d\lambda,$$

where φ is a function of λ satisfying $\varphi(\lambda_1) = \varphi_1$ and $\varphi(\lambda_2) = \varphi_2$. The shortest path or geodesic entails finding that function $\varphi(\lambda)$ which minimizes s_{12}. This is an exercise in the calculus of variations and the minimizing condition is given by the Beltrami identity,

$$L - \varphi'\frac{\partial L}{\partial \varphi'} = \text{const.}$$

Substituting for L and using Eqs. (1) gives

$$R\sin\alpha = \text{const.}$$

Clairaut (1735) found this relation, using a geometrical construction; a similar derivation is presented by Lyusternik (1964). Differentiating this relation gives

$$d\alpha = \sin\varphi\, d\lambda.$$

This, together with Eqs. (1), leads to a system of ordinary differential equations for a geodesic

$$\frac{d\varphi}{ds} = \frac{\cos\alpha}{\rho}; \quad \frac{d\lambda}{ds} = \frac{\sin\alpha}{\nu\cos\varphi}; \quad \frac{d\alpha}{ds} = \frac{\tan\varphi\sin\alpha}{\nu}.$$

We can express R in terms of the parametric latitude, β, using

$$R = a\cos\beta,$$

and Clairaut's relation then becomes

$$\sin\alpha_1\cos\beta_1 = \sin\alpha_2\cos\beta_2.$$

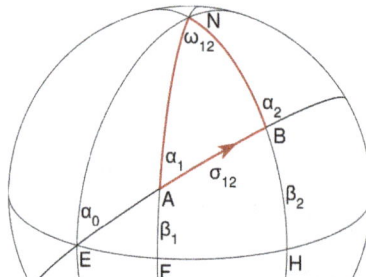

Fig. d. Geodesic problem mapped to the auxiliary sphere.

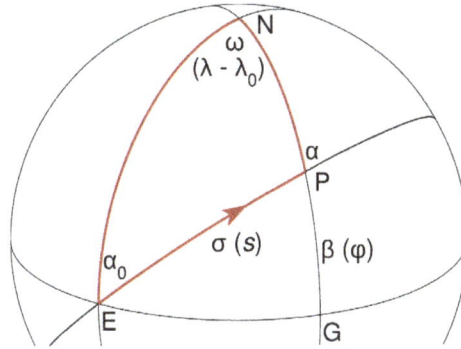

Fig. e. The elementary geodesic problem on the auxiliary sphere.

This is the sine rule of spherical trigonometry relating two sides of the triangle NAB, $NA = \frac{1}{2}\pi - \beta_1$ and $NB = \frac{1}{2}\pi - \beta_2$ and their opposite angles $B = \pi - \alpha_2$ and $A = \alpha_1$.

In order to find the relation for the third side $AB = \sigma_{12}$, the *spherical arc length*, and included angle $N = \omega_{12}$, the *spherical longitude*, it is useful to consider the triangle NEP representing a geodesic starting at the equator; In this figure, the variables referred to the auxiliary sphere are shown with the corresponding quantities for the ellipsoid shown in parentheses. Quantities without subscripts refer to the arbitrary point P; E, the point at which the geodesic crosses the equator in the northward direction, is used as the origin for σ, s and ω.

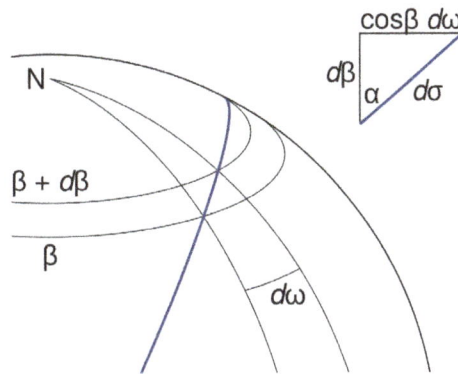

Fig. f. Differential element of a geodesic on a sphere.

If the side EP is extended by moving P infinitesimally, we obtain

$$(2) \qquad \cos\alpha\, d\sigma = d\beta, \quad \sin\alpha\, d\sigma = \cos\beta\, d\omega.$$

Combining Eqs. (1) and (2) gives differential equations for s and λ

$$\frac{1}{a}\frac{ds}{d\sigma} = \frac{d\lambda}{d\omega} = \frac{\sin\beta}{\sin\varphi}.$$

The relation between β and φ is

$$\tan \beta = \sqrt{1-e^2} \tan \varphi = (1-f) \tan \varphi,$$

which gives

$$\frac{\sin \beta}{\sin \varphi} = \sqrt{1-e^2 \cos^2 \beta},$$

so that the differential equations for the geodesic become

$$\frac{1}{a} \frac{ds}{d\sigma} = \frac{d\lambda}{d\omega} = \sqrt{1-e^2 \cos^2 \beta}.$$

The last step is to use σ as the independent parameter in both of these differential equations and thereby to express s and λ as integrals. Applying the sine rule to the vertices E and G in the spherical triangle EGP in Fig. e gives

$$\sin \beta = \sin \beta(\sigma; \alpha_0) = \cos \alpha_0 \sin \sigma,$$

where α_0 is the azimuth at E. Substituting this into the equation for $ds/d\sigma$ and integrating the result gives

$$(3) \qquad \frac{s}{b} = \int_0^{\sigma} \sqrt{1+k^2 \sin^2 \sigma'} \, d\sigma',$$

where

$$k = e' \cos \alpha_0,$$

and the limits on the integral are chosen so that $s(\sigma = 0) = 0$. Legendre (1811) pointed <u>out that the</u> equation for s is the same as the equation for the arc on an ellipse with semi-axes $b\sqrt{1 + e'^2 \cos^2 \alpha_0}$ and b. In order to express the equation for λ in terms of σ, we write

$$d\omega = \frac{\sin \alpha_0}{\cos^2 \beta} d\sigma,$$

which follows from Eq. (2) and Clairaut's relation. This yields

$$(4) \qquad \lambda - \lambda_0 = \omega - f \sin \alpha_0 \int_0^{\sigma} \frac{2-f}{1+(1-f)\sqrt{1+k^2 \sin^2 \sigma'}} d\sigma',$$

and the limits on the integrals are chosen so that $\lambda = \lambda_0$ at the equator crossing, $\sigma = 0$.

This completes the solution of the path of a geodesic using the auxiliary sphere. By this device a great circle can be mapped exactly to a geodesic on an ellipsoid of revolution.

There are also several ways of approximating geodesics on a terrestrial ellipsoid (with small flattening) (Rapp 1991). However, these are typically comparable in complexity to the method for the exact solution (Jekeli 2012).

Behavior of Geodesics

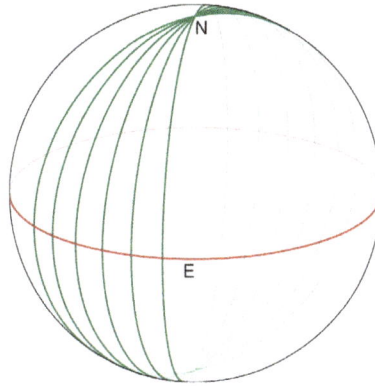

Fig. g. Meridians and the equator are the only closed geodesics.
(For the very flattened ellipsoids, there are other closed geodesics;

Geodesic on an oblate ellipsoid $\left(f = \frac{1}{50}\right)$ with $\alpha_0 = 45°$

Fig. h. Following the geodesic on the ellipsoid for about 5 circuits.	Fig. i. The same geodesic after about 70 circuits.	Fig. j. Geodesic on a prolate ellipsoid $(f = -\frac{1}{50})$ with $\alpha_0 = 45°$. Compare with Fig. h.

Fig.g shows the simple closed geodesics which consist of the meridians (green) and the equator (red). Here the qualification "simple" means that the geodesic closes on itself without an intervening self-intersection.

All other geodesics are typified by Figs.h and i which show a geodesic starting on the equator with $\alpha_0 = 45°$. The geodesic oscillates about the equator. The equatorial crossings are called *nodes* and the points of maximum or minimum latitude are called *vertices*; the vertex latitudes are given by $\beta = \pm(\frac{1}{2}\pi - |\alpha_0|)$. The geodesic completes one full oscillation in latitude before the longitude has increased by 360°. Thus, on each successive northward crossing of the equator, λ falls short of a full circuit of the equator by approximately $2\pi f \sin\alpha_0$ (for a prolate ellipsoid, this quantity is negative and λ completes more that a full circuit. For nearly all values of α_0, the geodesic will fill that portion of the ellipsoid between the two vertex latitudes.

Two additional closed geodesics for the oblate ellipsoid, $b/a = {}^2/_7$.

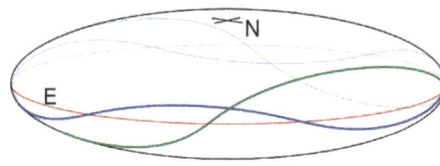

Fig. k. Side view.

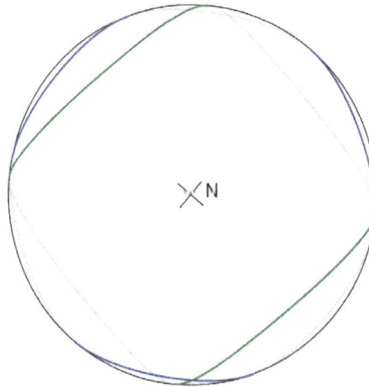

Fig. l. Top view.

If the ellipsoid is sufficiently oblate, i.e., $b/a < {}^1/_2$, another class of simple closed geodesics is possible (Klingenberg 1982). Two such geodesics are illustrated in Figs. k and l. Here $b/a = {}^2/_7$ and the equatorial azimuth, α_o, for the green (resp. blue) geodesic is chosen to be 53.175° (resp. 75.192°), so that the geodesic completes 2 (resp. 3) complete oscillations about the equator on one circuit of the ellipsoid.

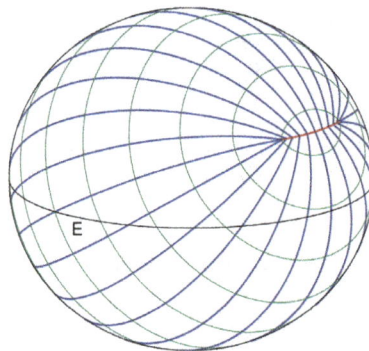

Fig. m. Geodesics (blue) from a single point for $f = {}^1/_{10}$, $\varphi_1 = -30°$; geodesic circles are shown in green and the cut locus in red.

Fig. m shows geodesics (in blue) emanating A with α_1 a multiple of 15° up to the point at which they cease to be shortest paths. (The flattening has been increased to $^1/_{10}$ in order to accentuate the ellipsoidal effects.) Also shown (in green) are curves of constant s_{12}, which are the geodesic circles centered A. Gauss (1828) showed that, on any surface, geodesics and geodesic circle intersect at right angles. The red line is the cut locus, the locus of points which have multiple (two in this case) shortest geodesics from A. On a sphere, the cut locus is a point. On an oblate ellipsoid (shown here), it is a segment of the circle of latitude centered on the point antipodal to A, $\varphi = -\varphi_1$. The longitudinal extent of cut locus is approximately $\lambda_{12} \in [\pi - f\pi \cos\varphi_1, \pi + f\pi \cos\varphi_1]$. If A lies on the equator, $\varphi_1 = 0$, this relation is exact and as a consequence the equator is only a shortest geodesic if $|\lambda_{12}| \le (1 - f)\pi$. For a prolate ellipsoid, the cut locus is a segment of the anti-meridian centered on the point antipodal to A, $\lambda_{12} = \pi$, and this means that meridional geodesics stop being shortest paths before the antipodal point is reached.

Solution of the Direct and Inverse Problems

Solving the geodesic problems entails mapping the geodesic onto the auxiliary sphere and solving the corresponding problem in great-circle navigation. When solving the "elementary" spherical triangle for NEP in Fig. e, Napier's rules for quadrantal triangles can be employed,

$$\sin \alpha_0 = \sin \alpha \cos \beta = \tan \omega \cot \sigma,$$
$$\cos \sigma = \cos \beta \cos \omega = \tan \alpha_0 \cot \alpha,$$
$$\cos \alpha = \cos \omega \cos \alpha_0 = \cot \sigma \tan \beta,$$
$$\sin \beta = \cos \alpha_0 \sin \sigma = \cot \alpha \tan \omega,$$
$$\sin \omega = \sin \sigma \sin \alpha = \tan \beta \tan \alpha_0.$$

The mapping of the geodesic involves evaluating the integrals for the distance, s, and the longitude, λ, Eqs. (3) and (4) and these depend on the parameter α_0.

Handling the direct problem is straightforward, because α_0 can be determined directly from the given quantities φ_1 and α_1.

In the case of the inverse problem, λ_{12} is given; this cannot be easily related to the equivalent spherical angle ω_{12} because α_0 is unknown. Thus, the solution of the problem requires that α_0 be found iteratively.

In geodetic applications, where f is small, the integrals are typically evaluated as a series (Legendre 1806) (Oriani 1806) (Bessel 1825) (Helmert 1880) (Rainsford 1955) (Rapp 1993). For arbitrary f, the integrals (3) and (4) can be found by numerical quadrature or by expressing them in terms of elliptic integrals (Legendre 1806) (Cayley 1870).

Vincenty (1975) provides solutions for the direct and inverse problems; these are based on a series expansion carried out to third order in the flattening and provide an accuracy of about 0.1 mm for the WGS84 ellipsoid; however the inverse method fails to converge for nearly antipodal points. Karney (2013) continues the expansions to sixth order which suffices to provide full double precision accuracy for $|f| \le \frac{1}{50}$ and improves the solution of the inverse problem so that it converges in all cases. Karney (2013, addendum) extends the method to use elliptic integrals which can be applied to ellipsoids with arbitrary flattening.

Differential Properties of Geodesics

Various problems involving geodesics require knowing their behavior when they are perturbed. This is useful in trigonometric adjustments (Ehlert 1993), determining the physical properties of signals which follow geodesics, etc. Consider a reference geodesic, parameterized by s, and a second geodesic a small distance $t(s)$ away from it. Gauss (1828) showed that $t(s)$ obeys the Gauss-Jacobi equation

$$\frac{d^2 t(s)}{ds^2} = K(s)t(s),$$

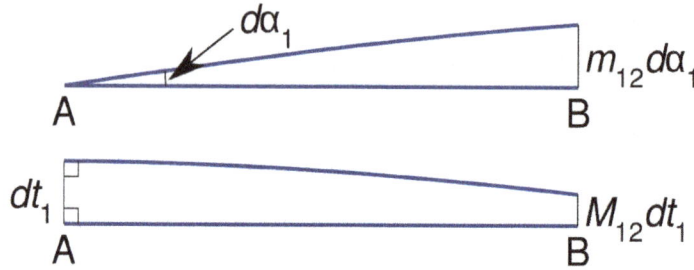

Fig. n. Definition of reduced length and geodesic scale.

where $K(s)$ is the Gaussian curvature at s. As a second order, linear, homogeneous differential equation, its solution may be expressed as the sum of two independent solutions

$$t(s_2) = Cm(s_1, s_2) + DM(s_1, s_2)$$

where

$$m(s_1, s_1) = 0, \quad \frac{dm(s_1, s_2)}{ds_2}\bigg|_{s_2 = s_1} = 1,$$

$$M(s_1, s_1) = 1, \quad \frac{dM(s_1, s_2)}{ds_2}\bigg|_{s_2 = s_1} = 0.$$

The quantity $m(s_1, s_2) = m_{12}$ is the so-called *reduced length*, and $M(s_1, s_2) = M_{12}$ is the *geodesic scale*. Their basic definitions are illustrated in Fig. n.

The Gaussian curvature for an ellipsoid of revolution is

$$K = \frac{1}{\rho \nu} = \frac{(1 - e^2 \sin^2 \varphi)^2}{b^2} = \frac{b^2}{a^4 (1 - e^2 \cos^2 \beta)^2}.$$

Helmert (1880, Eq. (6.5.1.)) solved the Gauss-Jacobi equation for this case enabling m_{12} and M_{12} to be expressed as integrals.

As we see from Fig. n (top sub-figure), the separation of two geodesics starting at the same point with azimuths differing by $d\alpha_1$ is $m_{12} d\alpha_1$. On a closed surface such as an ellipsoid, m_{12} oscillates about zero. The point at which m_{12} becomes zero is the point conjugate to the starting point. In order for a geodesic between A and B, of length s_{12}, to be a shortest path it must satisfy the Jacobi condition (Jacobi 1837) (Jacobi 1866) (Forsyth 1927) (Bliss 1916), that there is no point conjugate to A between A and B. If this condition is not satisfied, then there is a *nearby* path (not necessarily a geodesic) which is shorter. Thus, the Jacobi condition is a local property of the geodesic and is only a necessary condition for the geodesic being a global shortest path. Necessary and sufficient conditions for a geodesic being the shortest path are:

- for an oblate ellipsoid, $|\sigma_{12}| \leq \pi$;

- for a prolate ellipsoid, $|\lambda_{12}| \leq \pi$, if $\alpha_0 \neq 0$; if $\alpha_0 = 0$, the supplemental condition $m_{12} \geq 0$ is required if $|\lambda_{12}| = \pi$.

Envelope of Geodesics

Geodesics from a single point $\left(f = {}^1/_{10}, \varphi_1 = -30° \right)$

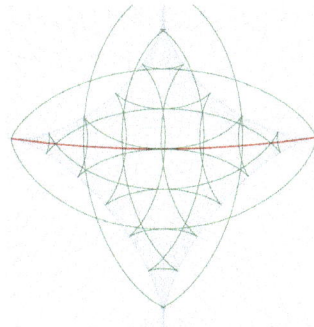

Fig. o. The envelope of geodesics from a point A at $\varphi_1 = -30°$.

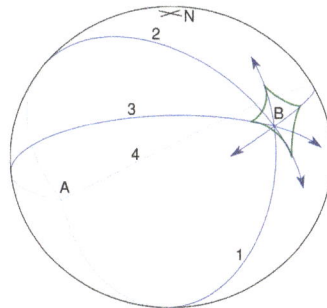

Fig. p. The four geodesics connecting A and a point B, $\varphi_2 = 26°$, $\lambda_{12} = 175°$.

The geodesics from a particular point A if continued past the cut locus form an envelope illustrated in Fig. o. Here the geodesics for which α_1 is a multiple of $3°$ are shown in light blue. (The geodesics are only shown for their first passage close to the antipodal point, not for subsequent ones.) Some geodesic circles are shown in green; these form cusps on the envelope. The cut locus is shown in red. The envelope is the locus of points which are conjugate to A; points on the envelope may be computed by finding the point at which $m_{12} = 0$ on a geodesic. Jacobi (1891) calls this star-like figure produced by the envelope an astroid.

Outside the astroid two geodesics intersect at each point; thus there are two geodesics (with a length approximately half the circumference of the ellipsoid) between A and these points. This corresponds to the situation on the sphere where there are "short" and "long" routes on a great circle between two points. Inside the astroid four geodesics intersect at each point. Four such geodesics are shown in Fig. p where the geodesics are numbered in order of increasing length. (This figure uses the same position for A as Fig. m and is drawn in the same projection.) The two shorter geodesics are *stable*, i.e., $m_{12} > 0$, so that there is no nearby path connecting the two points which is shorter; the other two are unstable. Only the shortest line (the first one) has $\sigma_{12} \leq \pi$. All the geodesics are tangent to the envelope which is shown in green in the figure.

The astroid is the (exterior) evolute of the geodesic circles centered at A. Likewise, the geodesic circles are involutes of the astroid.

Area of a Geodesic Polygon

A *geodesic polygon* is a polygon whose sides are geodesics. The area of such a polygon may be found by first computing the area between a geodesic segment and the equator, i.e., the area of the quadrilateral *AFHB* in Fig. a (Danielsen 1989). Once this area is known, the area of a polygon may be computed by summing the contributions from all the edges of the polygon.

Here an expression for the area S_{12} of *AFHB* is developed following Sjöberg (2006). The area of any closed region of the ellipsoid is

$$T = \int dT = \int \frac{1}{K} \cos \varphi \, d\varphi \, d\lambda,$$

where dT is an element of surface area and K is the Gaussian curvature. Now the Gauss–Bonnet theorem applied to a geodesic polygon states

$$\Gamma = \int K \, dT = \int \cos \varphi \, d\varphi \, d\lambda,$$

where

$$\Gamma = 2\pi - \sum_j \theta_j$$

is the geodesic excess and θ_j is the exterior angle at vertex j. Multiplying the equation for Γ by R_2^2, where R_2 is the authalic radius, and subtracting this from the equation for T gives

$$T = R_2^2 \Gamma + \int \left(\frac{1}{K} - R_2^2 \right) \cos \varphi \, d\varphi \, d\lambda$$

$$= R_2^2 \Gamma + \int \left(\frac{b^2}{(1 - e^2 \sin^2 \varphi)^2} - R_2^2 \right) \cos \varphi \, d\varphi \, d\lambda,$$

where the value of K for an ellipsoid has been substituted. Applying this formula to the quadrilateral *AFHB*, noting that $\Gamma = \alpha_2 - \alpha_1$, and performing the integral over φ gives

$$S_{12} = R_2^2 (\alpha_2 - \alpha_1) + b^2 \int_{\lambda_1}^{\lambda_2} \left(\frac{1}{2(1 - e^2 \sin^2 \varphi)} + \frac{\tanh^{-1}(e \sin \varphi)}{2 e \sin \varphi} - \frac{R_2^2}{b^2} \right) \sin \varphi \, d\lambda,$$

where the integral is over the geodesic line (so that φ is implicitly a function of λ). The integral can be expressed as a series valid for small f (Danielsen 1989) (Karney 2013).

The area of a geodesic polygon is given by summing S_{12} over its edges. This result holds provided that the polygon does not include a pole; if it does $2\pi R_2^2$ must be added to the sum. If the edges are specified by their vertices, then a convenient expression for te geodesic excess $E_{12} = \alpha_2 - \alpha_1$ is

$$\tan\frac{E_{12}}{2} = \frac{\sin\frac{1}{2}(\beta_2 + \beta_1)}{\cos\frac{1}{2}(\beta_2 - \beta_1)}\tan\frac{\omega_{12}}{2}.$$

Geodesics on a Triaxial Ellipsoid

Solving the geodesic problem for an ellipsoid of revolution is, from the mathematical point of view, relatively simple: because of symmetry, geodesics have a constant of the motion, given by Clairaut's relation allowing the problem to be reduced to quadrature. By the early 19th century (with the work of Legendre, Oriani, Bessel, et al.), there was a complete understanding of the properties of geodesics on an ellipsoid of revolution.

On the other hand, geodesics on a triaxial ellipsoid (with three unequal axes) have no obvious constant of the motion and thus represented a challenging "unsolved" problem in the first half of the 19th century. In a remarkable paper, Jacobi (1839) discovered a constant of the motion allowing this problem to be reduced to quadrature also (Klingenberg 1982).

The Triaxial Coordinate System

Consider the ellipsoid defined by

$$h = \frac{X^2}{a^2} + \frac{Y^2}{b^2} + \frac{Z^2}{c^2} = 1,$$

where (X, Y, Z) are Cartesian coordinates centered on the ellipsoid and, without loss of generality, $a \geq b \geq c > 0$. Jacobi (1866) employed the *ellipsoidal* latitude and longitude (β, ω) defined by

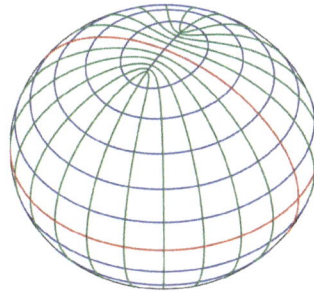

Fig. q. Ellipsoidal coordinates.

$$X = a\cos\omega\frac{\sqrt{a^2 - b^2\sin^2\beta - c^2\cos^2\beta}}{\sqrt{a^2 - c^2}},$$

$$Y = b\cos\beta\sin\omega,$$

$$Z = c\sin\beta\frac{\sqrt{a^2\sin^2\omega + b^2\cos^2\omega - c^2}}{\sqrt{a^2 - c^2}}.$$

In the limit $b \to a$, β becomes the parametric latitude for an oblate ellipsoid, so the use of the symbol β is consistent. However, ω is *different* from the spherical longitude defined above.

Grid lines of constant β (in blue) and ω (in green) are given in Fig. q. These constitute an orthogonal coordinate system: the grid lines intersect at right angles. The principal sections of the ellipsoid, defined by $X = 0$ and $Z = 0$ are shown in red. The third principal section, $Y = 0$, is covered by the lines $\beta = \pm 90°$ and $\omega = 0°$ or $\pm 180°$. These lines meet at four umbilical points (two of which are visible in this figure) where the principal radii of curvature are equal. Here and in the other figures the parameters of the ellipsoid are $a{:}b{:}c = 1.01{:}1{:}0.8$, and it is viewed in an orthographic projection from a point above $\varphi = 40°$, $\lambda = 30°$.

The grid lines of the ellipsoidal coordinates may be interpreted in three different ways:

1. They are "lines of curvature" on the ellipsoid: they are parallel to the directions of principal curvature (Monge 1796).

2. They are also intersections of the ellipsoid with confocal systems of hyperboloids of one and two sheets (Dupin 1813, Part 5).

3. Finally they are geodesic ellipses and hyperbolas defined using two adjacent umbilical points (Hilbert & Cohn-Vossen 1952, p. 188). For example, the lines of constant β in Fig. q can be generated with the familiar string construction for ellipses with the ends of the string pinned to the two umbilical points.

Jacobi's Solution

Jacobi showed that the geodesic equations, expressed in ellipsoidal coordinates, are separable. Here is how he recounted his discovery to his friend and neighbor Bessel (Jacobi 1839, Letter to Bessel),

The day before yesterday, I reduced to quadrature the problem of geodesic lines on an *ellipsoid with three unequal axes*. They are the simplest formulas in the world, Abelian integrals, which become the well known elliptic integrals if 2 axes are set equal.

The solution given by Jacobi (Jacobi 1839) (Jacobi 1866) is

$$\delta = \int \frac{\sqrt{b^2 \sin^2 \beta + c^2 \cos^2 \beta}\, d\beta}{\sqrt{a^2 - b^2 \sin^2 \beta - c^2 \cos^2 \beta}\sqrt{(b^2 - c^2)\cos^2 \beta - \gamma}}$$
$$- \int \frac{\sqrt{a^2 \sin^2 \omega + b^2 \cos^2 \omega}\, d\omega}{\sqrt{a^2 \sin^2 \omega + b^2 \cos^2 \omega - c^2}\sqrt{(a^2 - b^2)\sin^2 \omega + \gamma}}.$$

As Jacobi notes "a function of the angle β equals a function of the angle ω. These two functions are just Abelian integrals..." Two constants δ and γ appear in the solution. Typically δ is zero if the lower limits of the integrals are taken to be the starting point of the geodesic and the direction of the geodesics is determined by γ. However, for geodesics that start at an umbilical points, we have $\gamma = 0$ and δ determines the direction at the umbilical point. The constant γ may be expressed as

$$\gamma = (b^2 - c^2)\cos^2 \beta \sin^2 \alpha - (a^2 - b^2)\sin^2 \omega \cos^2 \alpha,$$

where α is the angle the geodesic makes with lines of constant ω. In the limit $b \to a$, this reduces to $\sin\alpha\cos\beta = \text{const.}$, the familiar Clairaut relation. A derivation of Jacobi's result is given by Darboux (1894); he gives the solution found by Liouville (1846) for general quadratic surfaces.

Survey of Triaxial Geodesics

Circumpolar geodesics, $\omega_1 = 0°$, $\alpha_1 = 90°$.

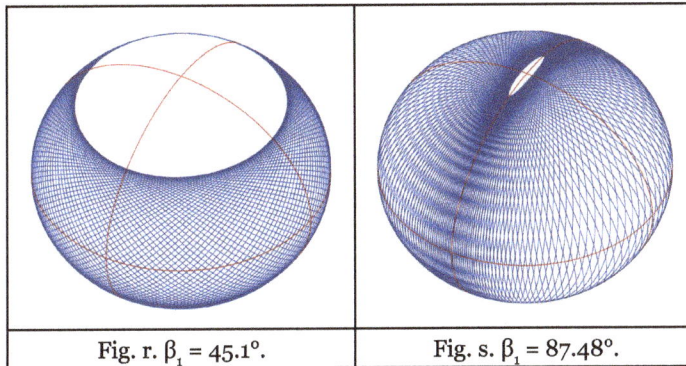

| Fig. r. $\beta_1 = 45.1°$. | Fig. s. $\beta_1 = 87.48°$. |

On a triaxial ellipsoid, there are only three simple closed geodesics, the three principal sections of the ellipsoid given by $X = 0$, $Y = 0$, and $Z = 0$. To survey the other geodesics, it is convenient to consider geodesics that intersect the middle principal section, $Y = 0$, at right angles. Such geodesics are shown in Figs. r–v, which use the same ellipsoid parameters and the same viewing direction as Fig. q. In addition, the three principal ellipses are shown in red in each of these figures.

If the starting point is $\beta_1 \in (-90°, 90°)$, $\omega_1 = 0$, and $\alpha_1 = 90°$, then $\gamma > 0$ and the geodesic encircles the ellipsoid in a "circumpolar" sense. The geodesic oscillates north and south of the equator; on each oscillation it completes slightly less than a full circuit around the ellipsoid resulting, in the typical case, in the geodesic filling the area bounded by the two latitude lines $\beta = \pm\beta_1$. Two examples are given in Figs. r and s. Figure r shows practically the same behavior as for an oblate ellipsoid of revolution (because $a \approx b$); compare to Fig. i. However, if the starting point is at a higher latitude (Fig. r) the distortions resulting from $a \neq b$ are evident. All tangents to a circumpolar geodesic touch the confocal single-sheeted hyperboloid which intersects the ellipsoid at $\beta = \beta_1$ (Chasles 1846) (Hilbert & Cohn-Vossen 1952).

Transpolar geodesics, $\beta_1 = 90°$, $\alpha_1 = 180°$.

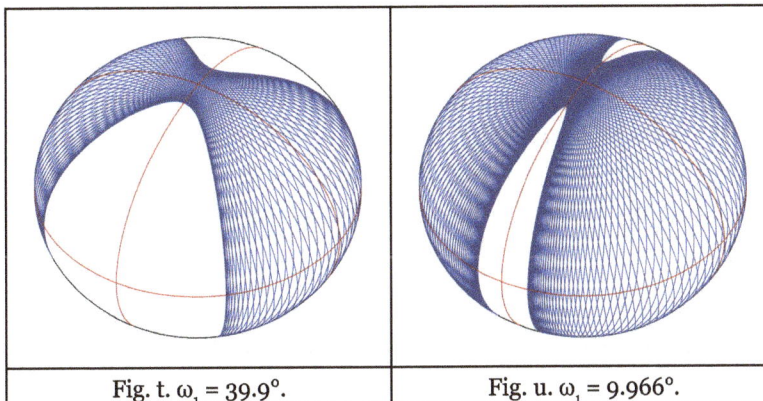

| Fig. t. $\omega_1 = 39.9°$. | Fig. u. $\omega_1 = 9.966°$. |

If the starting point is $\beta_1 = 90°$, $\omega_1 \in (0°, 180°)$, and $\alpha_1 = 180°$, then $\gamma < 0$ and the geodesic encircles the ellipsoid in a "transpolar" sense. The geodesic oscillates east and west of the ellipse $X = 0$; on each oscillation it completes slightly more than a full circuit around the ellipsoid. In the typical case, this results in the geodesic filling the area bounded by the two longitude lines $\omega = \omega_1$ and $\omega = 180° - \omega_1$. If $a = b$, all meridians are geodesics; the effect of $a \neq b$ causes such geodesics to oscillate east and west. Two examples are given in Figs. t and u. The constriction of the geodesic near the pole disappears in the limit $b \to c$; in this case, the ellipsoid becomes a prolate ellipsoid and Fig. t would resemble Fig. j (rotated on its side). All tangents to a transpolar geodesic touch the confocal double-sheeted hyperboloid which intersects the ellipsoid at $\omega = \omega_1$.

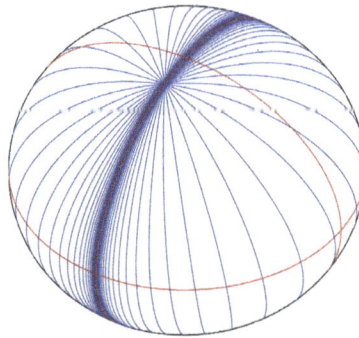

Fig. v. An umbilical geodesic, $\beta_1 = 90°$, $\omega_1 = 0°$, $\alpha_1 = 135°$.

If the starting point is $\beta_1 = 90°$, $\omega_1 = 0°$ (an umbilical point), and $\alpha_1 = 135°$ (the geodesic leaves the ellipse $Y = 0$ at right angles), then $\gamma = 0$ and the geodesic repeatedly intersects the opposite umbilical point and returns to its starting point. However, on each circuit the angle at which it intersects $Y = 0$ becomes closer to $0°$ or $180°$ so that asymptotically the geodesic lies on the ellipse $Y = 0$ (Hart 1849) (Arnold 1989), as shown in Fig. v. A single geodesic does not fill an area on the ellipsoid. All tangents to umbilical geodesics touch the confocal hyperbola that intersects the ellipsoid at the umbilic points.

Umbilical geodesic enjoy several interesting properties.

- Through any point on the ellipsoid, there are two umbilical geodesics.

- The geodesic distance between opposite umbilical points is the same regardless of the initial direction of the geodesic.

- Whereas the closed geodesics on the ellipses $X = 0$ and $Z = 0$ are stable (an geodesic initially close to and nearly parallel to the ellipse remains close to the ellipse), the closed geodesic on the ellipse $Y = 0$, which goes through all 4 umbilical points, is *exponentially unstable*. If it is perturbed, it will swing out of the plane $Y = 0$ and flip around before returning to close to the plane. (This behavior may repeat depending on the nature of the initial perturbation.)

If the starting point A of a geodesic is not an umbilical point, its envelope is an astroid with two cusps lying on $\beta = -\beta_1$ and the other two on $\omega = \omega_1 + \pi$. The cut locus for A is the portion of the line $\beta = -\beta_1$ between the cusps.

Applications

The direct and inverse geodesic problems no longer play the central role in geodesy that they once did. Instead of solving adjustment of geodetic networks as a two-dimensional problem in spheroidal trigonometry, these problems are now solved by three-dimensional methods (Vincenty & Bowring 1978). Nevertheless, terrestrial geodesics still play an important role in several areas:

- for measuring distances and areas in geographic information systems;

- the definition of maritime boundaries (UNCLOS 2006);

- in the rules of the Federal Aviation Administration for area navigation (RNAV 2007);

- the method of measuring distances in the FAI Sporting Code (FAI 2013).

By the principle of least action, many problems in physics can be formulated as a variational problem similar to that for geodesics. Indeed, the geodesic problem is equivalent to the motion of a particle constrained to move on the surface, but otherwise subject to no forces (Laplace 1799a) (Hilbert & Cohn-Vossen 1952,). For this reason, geodesics on simple surfaces such as ellipsoids of revolution or triaxial ellipsoids are frequently used as "test cases" for exploring new methods. Examples include:

- the development of elliptic integrals (Legendre 1811) and elliptic functions (Weierstrass 1861);

- the development of differential geometry (Gauss 1828) (Christoffel 1869);

- methods for solving systems of differential equations by a change of independent variables (Jacobi 1839);

- the study of caustics (Jacobi 1891);

- investigations into the number and stability of periodic orbits (Poincaré 1905);

- in the limit $c \to 0$, geodesics on a triaxial ellipsoid reduce to a case of dynamical billiards;

- extensions to an arbitrary number of dimensions (Knörrer 1980);

- geodesic flow on a surface (Berger 2010, Chap. 12).

Earth Ellipsoid

An Earth ellipsoid is a mathematical figure approximating the shape of the Earth, used as a reference frame for computations in geodesy, astronomy and the geosciences. Various different ellipsoids have been used as approximations.

It is an ellipsoid of revolution, whose short (polar) axis (connecting the two flattest spots called geographical north and south poles) is approximately aligned with the rotation axis of the Earth. The ellipsoid is defined by the equatorial axis a and the polar axis b; their difference is about 21 km or 0.335 percent. Additional parameters are the mass function $J2$, the correspondent gravity formula, and the rotation period (usually 86164 seconds).

Many methods exist for determination of the axes of an Earth ellipsoid, ranging from meridian arcs up to modern satellite geodesy or the analysis and interconnection of continental geodetic networks. Amongst the different set of data used in national surveys are several of special importance: the Bessel ellipsoid of 1841, the international Hayford ellipsoid of 1924, and (for GPS positioning) the WGS84 ellipsoid.

Mean Earth Ellipsoid and Reference Ellipsoids

A data set which describes the global average of the Earth's surface curvature is called the *mean* Earth Ellipsoid. It refers to a theoretical coherence between the geographic latitude and the meridional curvature of the geoid. The latter is close to the mean sea level, and therefore an ideal Earth ellipsoid has the same volume as the geoid.

While the *mean* Earth ellipsoid is the ideal basis of global geodesy, for regional networks a so-called *reference ellipsoid* may be the better choice. When geodetic measurements have to be computed on a mathematical reference surface, this surface should have a similar curvature as the regional geoid - otherwise, reduction of the measurements will get small distortions.

This is the reason for the "long life" of former reference ellipsoids like the Hayford or the Bessel ellipsoid, despite the fact that their main axes deviate by several hundred meters from the modern values. Another reason is a judicial one: the coordinates of millions of boundary stones should remain fixed for a long period. If their reference surface changes, the coordinates themselves also change.

However, for international networks, GPS positioning, or astronautics, these regional reasons are less relevant. As knowledge of the Earth's figure is increasingly accurate, the International Geoscientific Union IUGG usually adapts the axes of the Earth ellipsoid to the best available data.

Historical Method Of Determining the Ellipsoid

High precision land surveys can be used to determine the distance between two places at nearly the same longitude by measuring a base line and a chain of triangles. (Suitable stations for the end points are rarely at the same longitude). The distance Δ along the meridian from one end point to a point at the same latitude as the second end point is then calculated by trigonometry. The surface distance Δ is reduced to Δ', the corresponding distance at mean sea level. The intermediate distances to points on the meridian at the same latitudes as other stations of the survey may also be calculated.

The geographic latitudes of both end points, φ_s (standpoint) and φ_f (forepoint) and possibly at other points are determined by astrogeodesy, observing the zenith distances of sufficient numbers of stars. If latitudes are measured at end points only, the radius of curvature at the midpoint of the meridian arc can be calculated from $R = \Delta'/(|\varphi_s\text{-}\varphi_f|)$. A second meridian arc will allow the derivation of two parameters required to specify a reference ellipsoid. Longer arcs with intermediate latitude determinations can completely determine the ellipsoid. In practice multiple arc measurements are used to determine the ellipsoid parameters by the method of least squares. The parameters determined are usually the semi-major axis,   , and either the semi-minor axis, b, or the inverse flattening $1/f$, (where the flattening is $f = (a-b)/a$).

Geodesy no longer uses simple meridian arcs, but complex networks with hundreds of fixed points linked by the methods of satellite geodesy.

Historical Earth Ellipsoids

The reference ellipsoid models listed below have had utility in geodetic work and many are still in use. The older ellipsoids are named for the individual who derived them and the year of development is given. In 1887 the English surveyor Colonel Alexander Ross Clarke CB FRS RE was awarded the Gold Medal of the Royal Society for his work in determining the figure of the Earth. The international ellipsoid was developed by John Fillmore Hayford in 1910 and adopted by the International Union of Geodesy and Geophysics (IUGG) in 1924, which recommended it for international use.

At the 1967 meeting of the IUGG held in Lucerne, Switzerland, the ellipsoid called GRS-67 (Geodetic Reference System 1967) in the listing was recommended for adoption. The new ellipsoid was not recommended to replace the International Ellipsoid (1924), but was advocated for use where a greater degree of accuracy is required. It became a part of the GRS-67 which was approved and adopted at the 1971 meeting of the IUGG held in Moscow. It is used in Australia for the Australian Geodetic Datum and in South America for the South American Datum 1969.

The GRS-80 (Geodetic Reference System 1980) as approved and adopted by the IUGG at its Canberra, Australia meeting of 1979 is based on the equatorial radius (semi-major axis of Earth ellipsoid) a, total mass GM, dynamic form factor J_2 and angular velocity of rotation ω, making the inverse flattening $1/f$ a derived quantity. The minute difference in $1/f$ seen between GRS-80 and WGS-84 results from an unintentional truncation in the latter's defining constants: while the WGS-84 was designed to adhere closely to the GRS-80, incidentally the WGS-84 derived flattening turned out to be slightly different than the GRS-80 flattening because the normalized second degree zonal harmonic gravitational coefficient, that was derived from the GRS-80 value for J2, was truncated to 8 significant digits in the normalization process.

An ellipsoidal model describes only the ellipsoid's geometry and a normal gravity field formula to go with it. Commonly an ellipsoidal model is part of a more encompassing geodetic datum. For example, the older ED-50 (European Datum 1950) is based on the Hayford or International Ellipsoid. WGS-84 is peculiar in that the same name is used for both the complete geodetic reference system and its component ellipsoidal model. Nevertheless, the two concepts—ellipsoidal model and geodetic reference system—remain distinct.

Reference Ellipsoid

In geodesy, a reference ellipsoid is a mathematically defined surface that approximates the geoid, the truer figure of the Earth, or other planetary body. Because of their relative simplicity, reference ellipsoids are used as a preferred surface on which geodetic network computations are performed and point coordinates such as latitude, longitude, and elevation are defined.

Ellipsoid Parameters

In 1687 Isaac Newton published the Principia in which he included a proof that a rotating self-gravitating fluid body in equilibrium takes the form of an oblate ellipsoid of revolution which he termed an oblate spheroid. Current practice uses the word 'ellipsoid' alone in preference to the full term 'oblate ellipsoid of revolution' or the older term 'oblate spheroid'. In the rare instances (some asteroids and planets) where a more general ellipsoid shape is required as a model the term used is triaxial (or scalene) ellipsoid. A great many ellipsoids have been used with various sizes and centres but modern (post-GPS) ellipsoids are centred at the actual center of mass of the Earth or body being modeled.

The shape of an (oblate) ellipsoid (of revolution) is determined by the shape parameters of that ellipse which generates the ellipsoid when it is rotated about its minor axis. The semi-major axis of the ellipse, a, is identified as the equatorial radius of the ellipsoid: the semi-minor axis of the ellipse, b, is identified with the polar distances (from the centre). These two lengths completely specify the shape of the ellipsoid but in practice geodesy publications classify reference ellipsoids by giving the semi-major axis and the *inverse* flattening, $\frac{1}{f}$, The flattening, f, is simply a measure of how much the symmetry axis is compressed relative to the equatorial radius:

$$f = \frac{a - b}{a}.$$

For the Earth, f is around $\frac{1}{300}$ corresponding to a difference of the major and minor semi-axes of approximately 21 km (13 miles). Some precise values are given in the table below and also in Figure of the Earth. For comparison, Earth's Moon is even less elliptical, with a flattening of less than $\frac{1}{825}$, while Jupiter is visibly oblate at about $\frac{1}{15}$ and one of Saturn's triaxial moons, Telesto, is nearly $\frac{1}{3}$ to $\frac{1}{2}$.

A great many other parameters are used in geodesy but they can all be related to one or two of the set a, b and f. They are listed in ellipse.

Coordinates

A primary use of reference ellipsoids is to serve as a basis for a coordinate system of latitude (north/south), longitude (east/west), and elevation (height). For this purpose it is necessary to identify a *zero meridian*, which for Earth is usually the Prime Meridian. For other bodies a fixed surface feature is usually referenced, which for Mars is the meridian passing through the crater Airy-0. It is possible for many different coordinate systems to be defined upon the same reference ellipsoid.

The longitude measures the rotational angle between the zero meridian and the measured point. By convention for the Earth, Moon, and Sun it is expressed in degrees ranging from −180° to +180° For other bodies a range of 0° to 360° is used.

The latitude measures how close to the poles or equator a point is along a meridian, and is represented as an angle from −90° to +90°, where 0° is the equator. The common or *geodetic latitude* is the angle between the equatorial plane and a line that is normal to the reference ellipsoid. Depending on the flattening, it may be slightly different from the *geocentric (geographic) latitude*, which is the angle between the equatorial plane and a line from the center of the ellipsoid. For non-Earth bodies the terms *planetographic* and *planetocentric* are used instead.

The coordinates of a geodetic point are customarily stated as geodetic latitude and longitude, i.e., the direction in space of the geodetic normal containing the point, and the height h of the point over the reference ellipsoid.. If these coordinates, i.e., latitude ϕ, longitude λ and height h, are given, one can compute the *geocentric rectangular coordinates* of the point as follows:

$$X = \left(N(\phi) + h\right)\cos\phi\cos\lambda$$
$$Y = \left(N(\phi) + h\right)\cos\phi\sin\lambda$$
$$Z = \left(\frac{b^2}{a^2}N(\phi) + h\right)\sin\phi$$

where

$$N(\phi) = \frac{a^2}{\sqrt{a^2\cos^2\phi + b^2\sin^2\phi}},$$

and a and b are the equatorial radius (semi-major axis) and the polar radius (semi-minor axis), respectively. N is the *radius of curvature in the prime vertical.*

In contrast, extracting φ, λ and h from the rectangular coordinates usually requires iteration. A straightforward method is given in an OSGB publication and also in web notes. More sophisticated methods are outlined in geodetic system.

Historical Earth Ellipsoids

Currently the most common reference ellipsoid used, and that used in the context of the Global Positioning System, is the one defined by WGS 84.

Traditional reference ellipsoids or *geodetic datums* are defined regionally and therefore non-geocentric, e.g., ED50. Modern geodetic datums are established with the aid of GPS and will therefore be geocentric, e.g., WGS 84.

Ellipsoids for other Planetary Bodies

Reference ellipsoids are also useful for geodetic mapping of other planetary bodies including planets, their satellites, asteroids and comet nuclei. Some well observed bodies such as the Moon and Mars now have quite precise reference ellipsoids.

For rigid-surface nearly-spherical bodies, which includes all the rocky planets and many moons, ellipsoids are defined in terms of the axis of rotation and the mean surface height excluding any atmosphere. Mars is actually egg shaped, where its north and south polar radii differ by approximately 6 km (4 miles), however this difference is small enough that the average polar radius is used to define its ellipsoid. The Earth's Moon is effectively spherical, having almost no bulge at its equator. Where possible a fixed observable surface feature is used when defining a reference meridian.

For gaseous planets like Jupiter, an effective surface for an ellipsoid is chosen as the equal-pressure boundary of one bar. Since they have no permanent observable features the choices of prime meridians are made according to mathematical rules.

Small moons, asteroids, and comet nuclei frequently have irregular shapes. For some of these, such as Jupiter's Io, a scalene (triaxial) ellipsoid is a better fit than the oblate spheroid. For highly irregular bodies the concept of a reference ellipsoid may have no useful value, so sometimes a spherical reference is used instead and points identified by planetocentric latitude and longitude. Even that can be problematic for non-convex bodies, such as Eros, in that latitude and longitude don't always uniquely identify a single surface location.

References

- Chasles, M. (1846). "Sur les lignes géodésiques et les lignes de courbure des surfaces du second degré" [Geodesic lines and the lines of curvature of the surfaces of the second degree]. Journal de Mathématiques Pures et Appliquées (in French). 11: 5–20

- Arnold, V. I. (1989). Mathematical Methods of Classical Mechanics. Translated by Vogtmann, K.; Weinstein, A. (2nd ed.). Springer-Verlag. ISBN 978-0-387-96890-2. OCLC 4037141

- Sjöberg, L. E. (2006). "Determination of areas on the plane, sphere and ellipsoid". Survey Review. 38 (301): 583–593. doi:10.1179/003962606780732100

- Hart, A. S. (1849). "Geometrical demonstration of some properties of geodesic lines". Cambridge and Dublin Mathematical Journal. 4: 80–84

- Bliss, G. A. (1916). "Jacobi's condition for problems of the calculus of variations in parametric form". Transactions of the American Mathematical Society. 17 (2): 195–206. doi:10.1090/S0002-9947-1916-1501037-4 (free access)

- Berger, M. (2010). Geometry Revealed. Translated by Senechal, L. J. Springer. ISBN 978-3-540-70996-1. doi:10.1007/978-3-540-70997-8

- Jacobi, C. G. J. (1837). "Zur Theorie der Variations-Rechnung und der Differential-Gleichungen" [The theory of the calculus of variations and of differential equations]. Journal für die reine und angewandte Mathematik (Crelles Journal) (in German). 1837 (17): 68–82. doi:10.1515/crll.1837.17.68

- Klingenberg, W. P. A. (1982). Riemannian Geometry. de Gruyer. ISBN 978-3-11-008673-7. MR 666697. OCLC 8476832

- Alexander, J. C. (1985). "The Numerics of Computing Geodetic Ellipsoids". SIAM Review. 27 (2): 241. doi:10.1137/1027056

- Borre, K.; Strang, W. G. (2012). Algorithms for Global Positioning. Wellesley-Cambridge Press. ISBN 978-0-9802327-3-8. OCLC 795014501. Chapter 11, Geometry of the Ellipsoid

Permissions

Index

www.ingramcontent.com/pod-product-compliance
Lightning Source LLC
Chambersburg PA
CBHW061246190326
41458CB00011B/3591